001

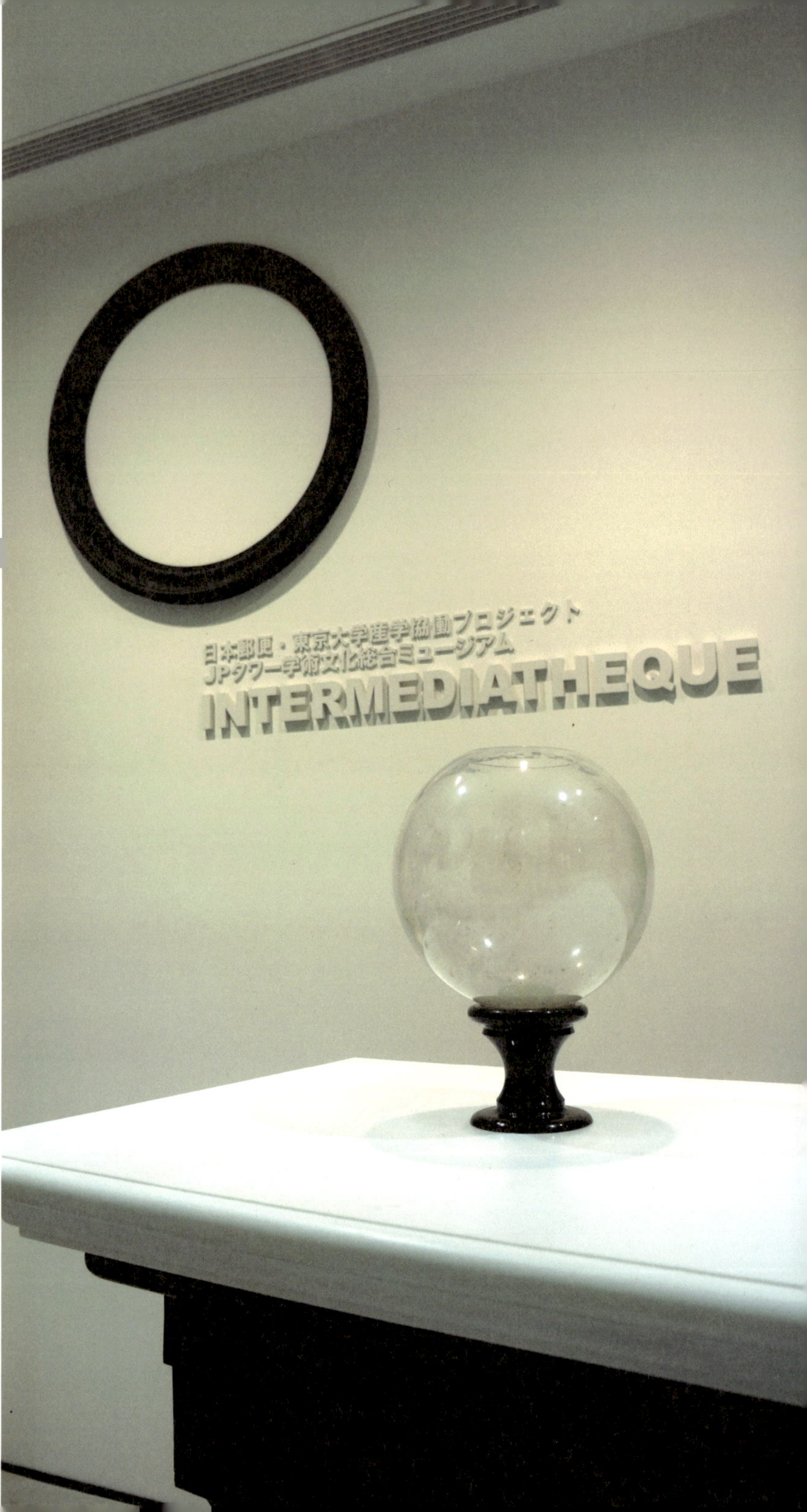
日本郵便・東京大学産学協働プロジェクト
JPタワー学術文化総合ミュージアム
INTERMEDIATHEQUE

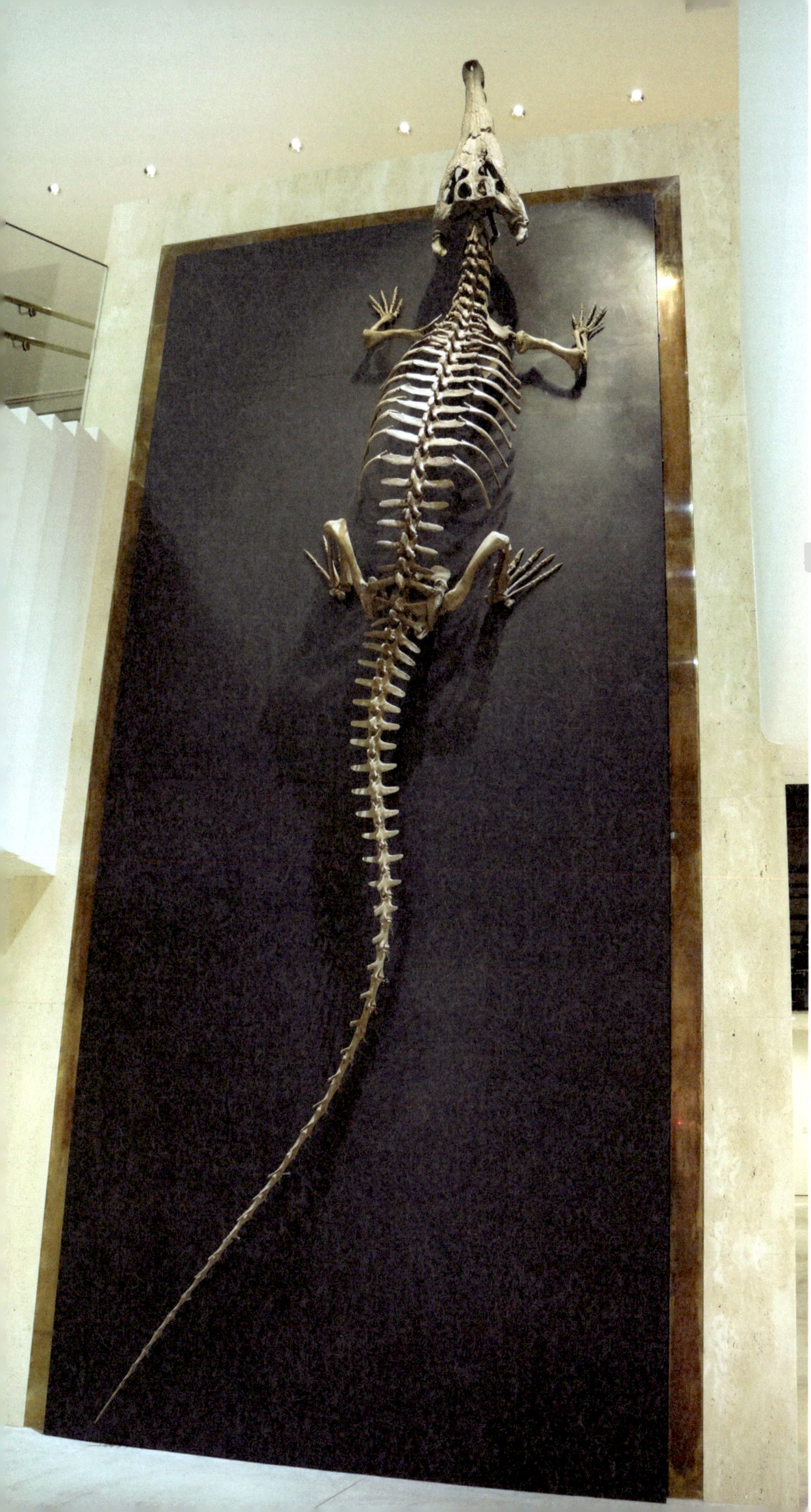

008

COLONNADE2

スッポン
Soft-Shelled Turtle

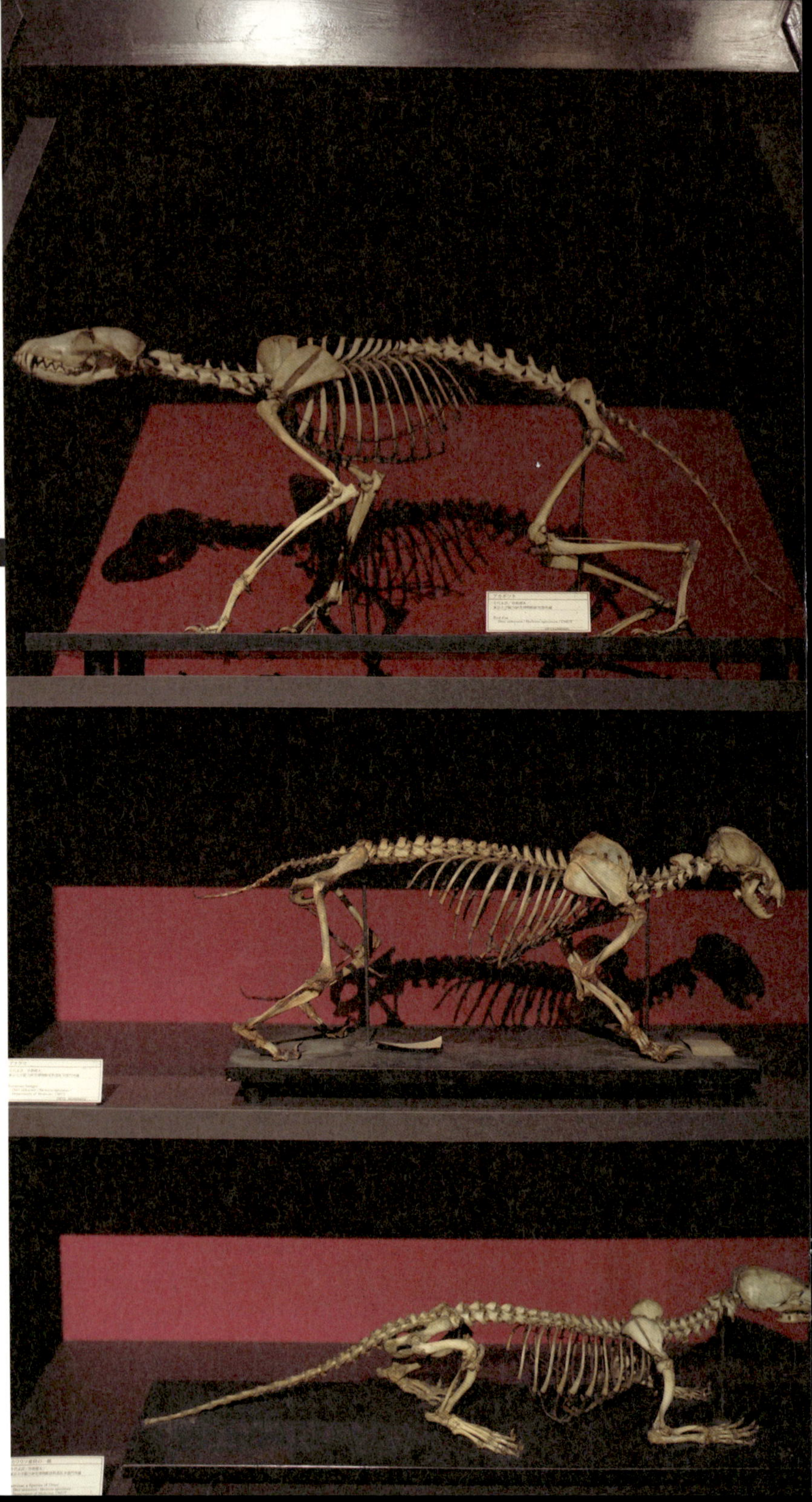

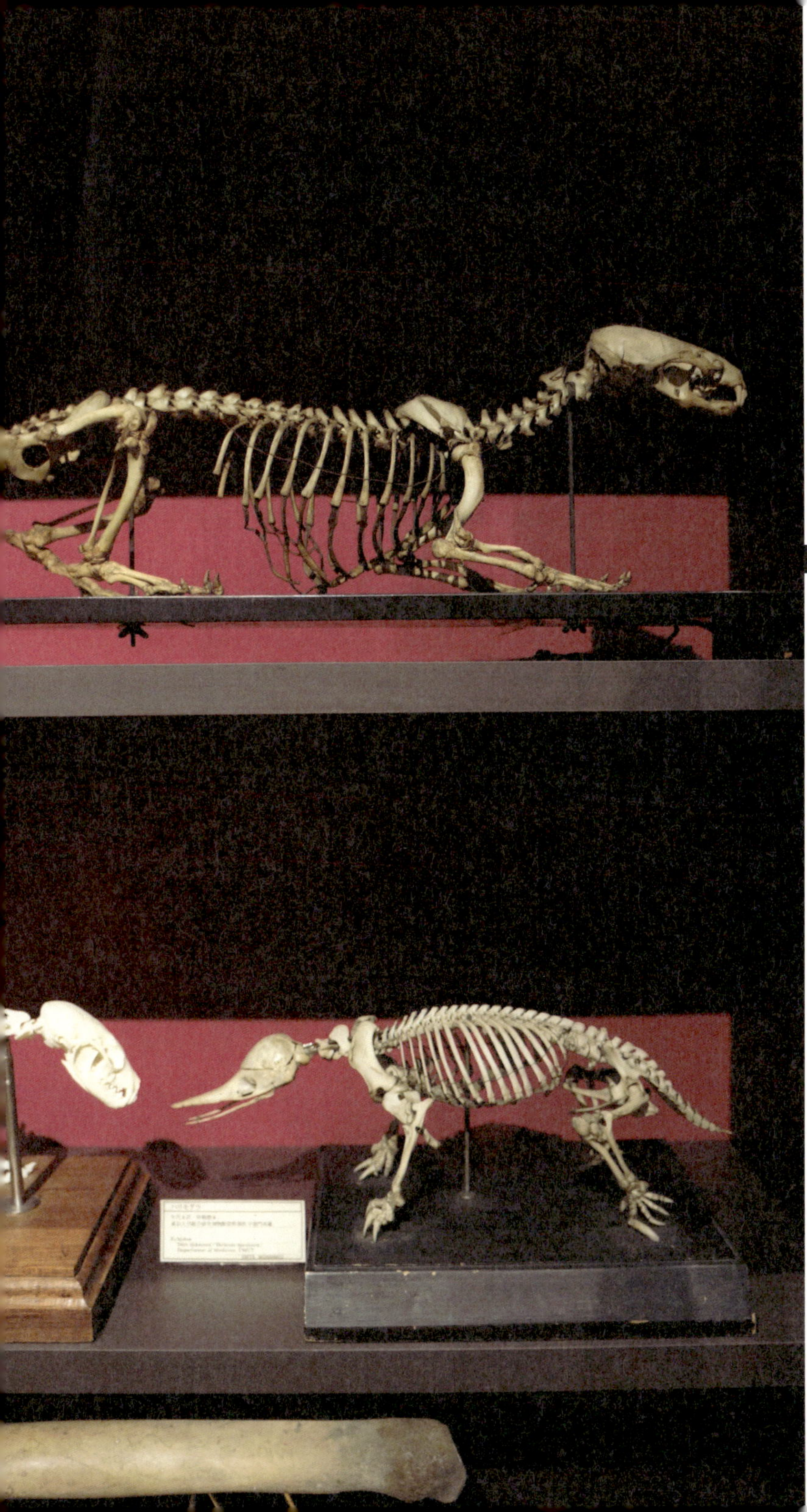

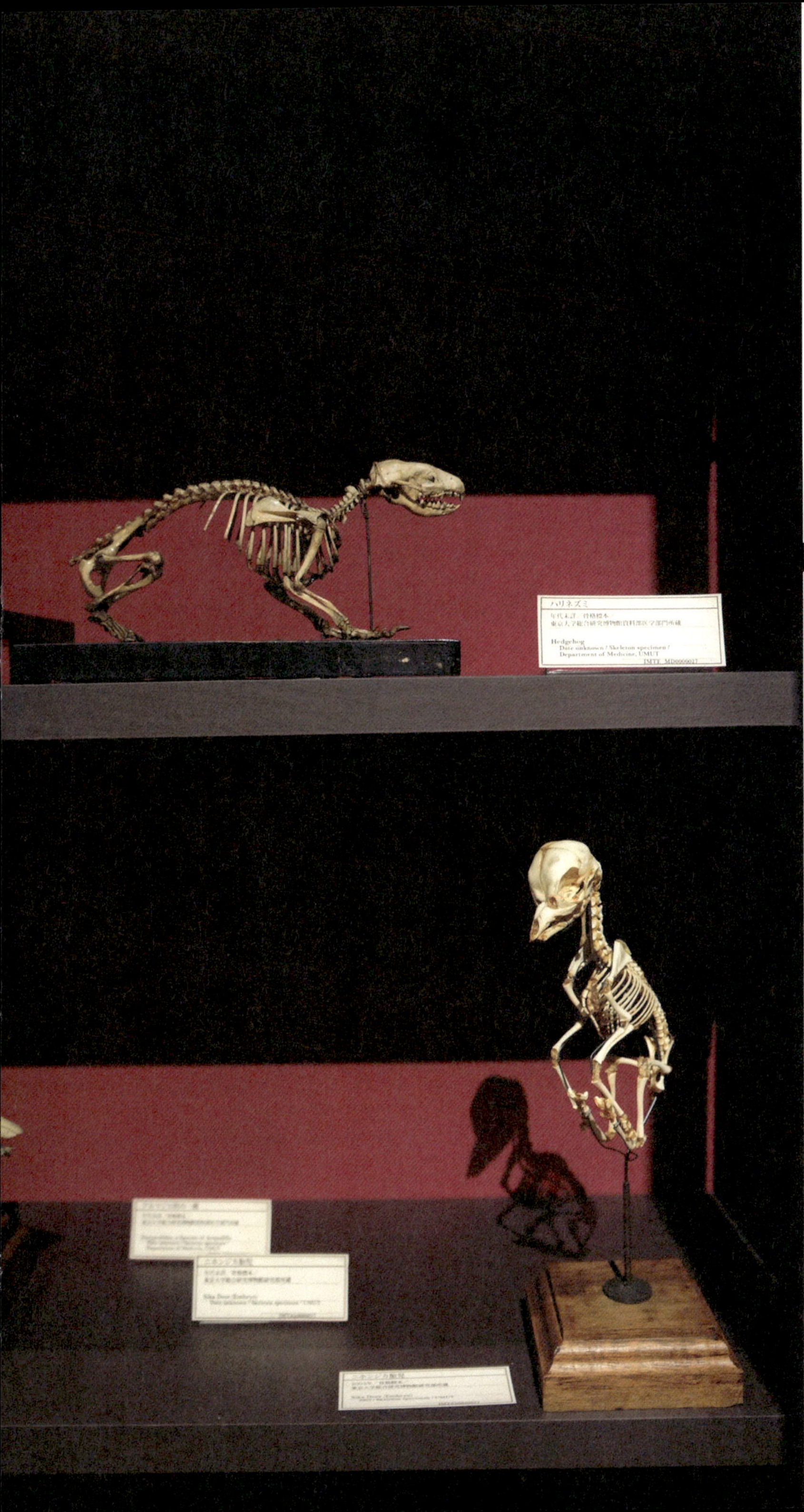
ハリネズミ
年代未詳／骨格標本／
東京大学総合研究博物館資料部医学部門所蔵
Hedgehog
Date unknown / Skeleton specimen /
Department of Medicine, UMUT

イエネコ
年代未詳／骨格標本／東京大学総合研究博物館研究部所蔵
Cat
Date unknown / Skeleton specimen / UMUT

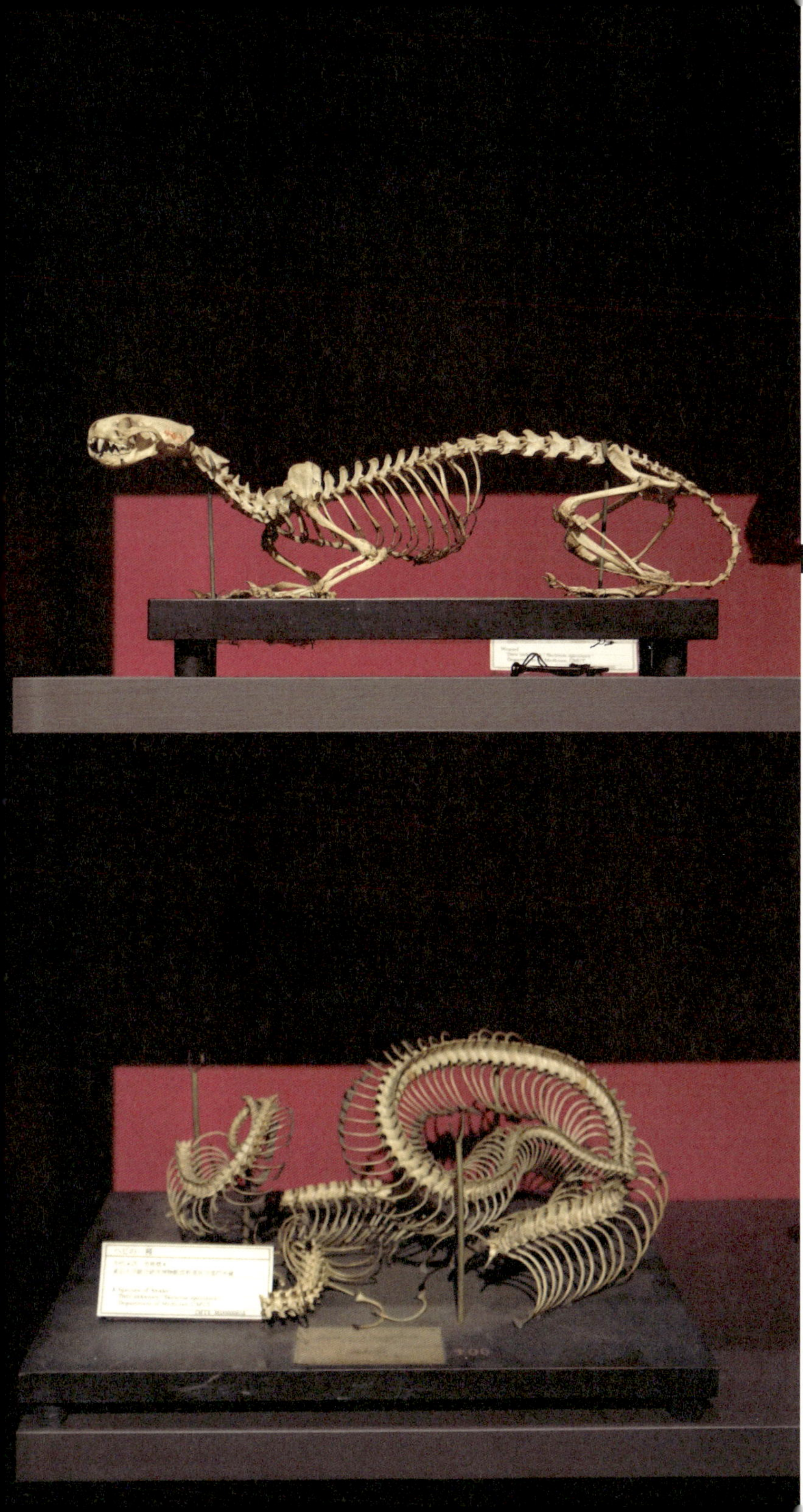

スズキ
年代未詳／骨格標本／
東京大学総合研究博物館資料部医学部門所蔵
Japanese Seabass
Date unknown / Skeleton specimen /
Department of Medicine, UMUT

ホタルジャコ科の一種
Department of Medicine, UMUT
843

Balistidae, a Species of Triggerfish
Date unknown / Skeleton specimen /
Department of Medicine, UMUT

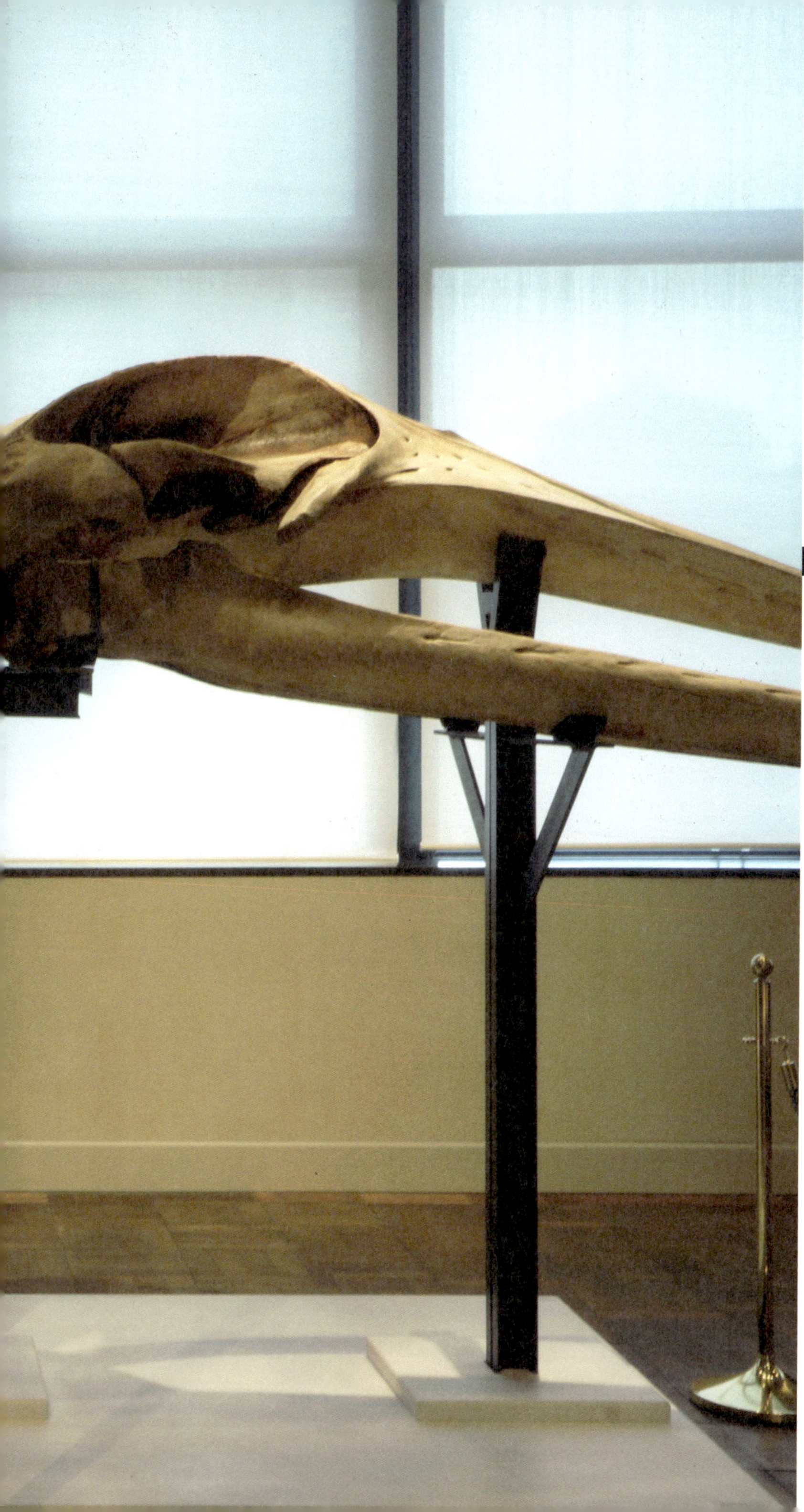

ニホンザル
IMTAb_UT0000171
2006年／骨格標本／
東京大学総合研究博物館研究部所蔵
Japanese Macaque
2006 / Skeleton specimen / UMUT

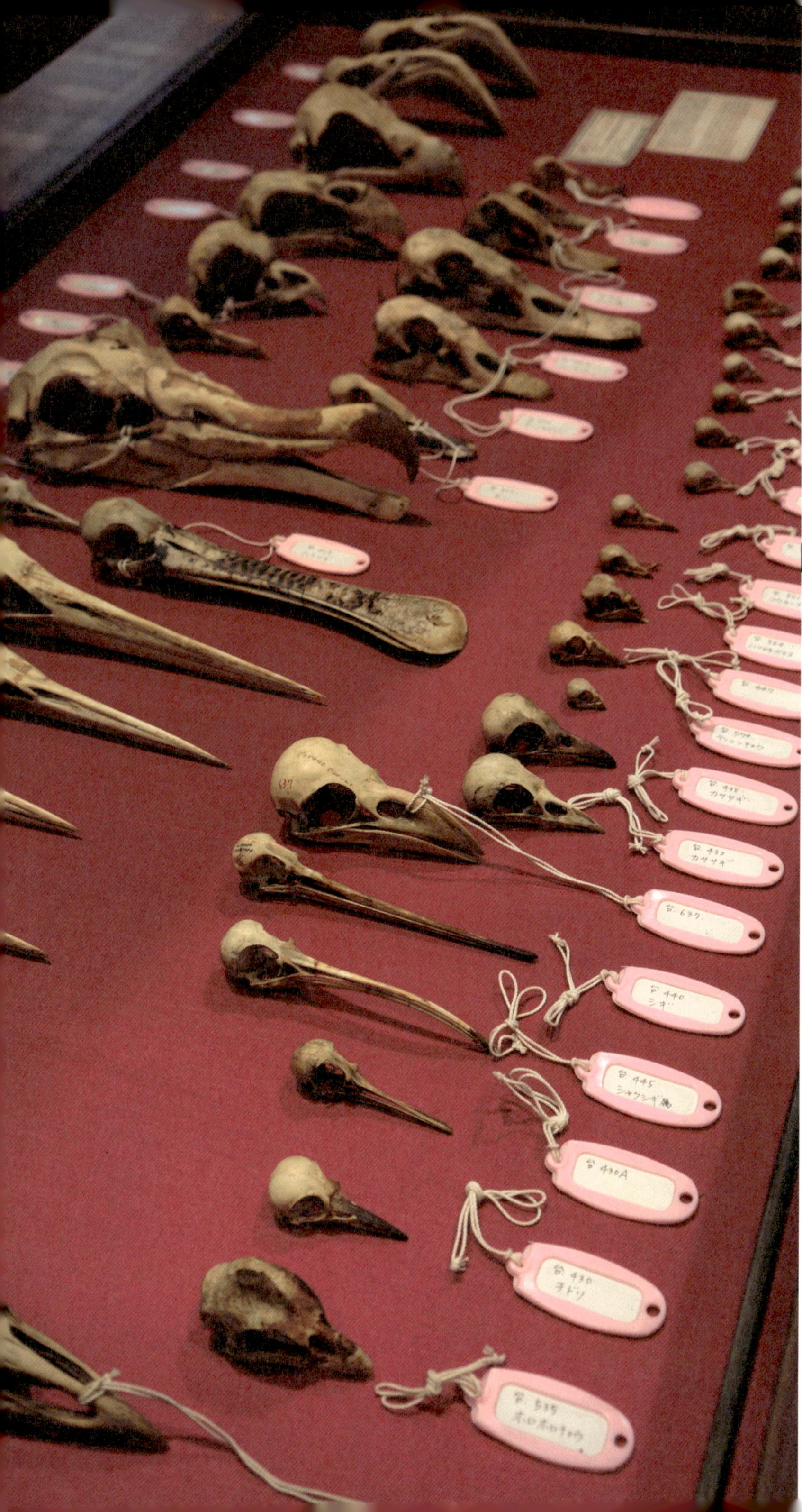
台 535
ホロホロチョウ
台 430
チドリ
台 430A
台 445
シャクシギ属
台 440
シギ
台.637.
台 430
カササギ

Japanese Macaque

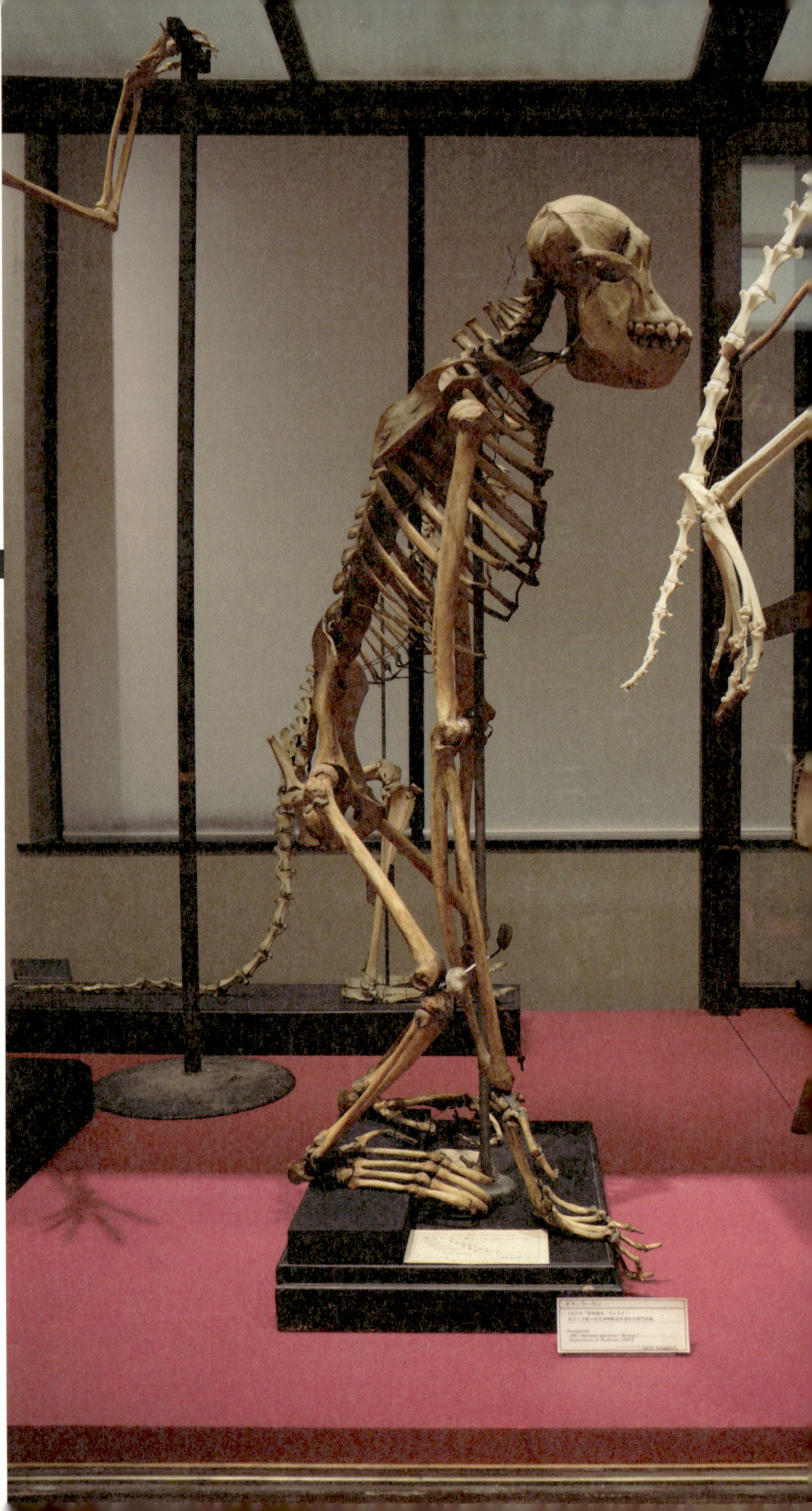

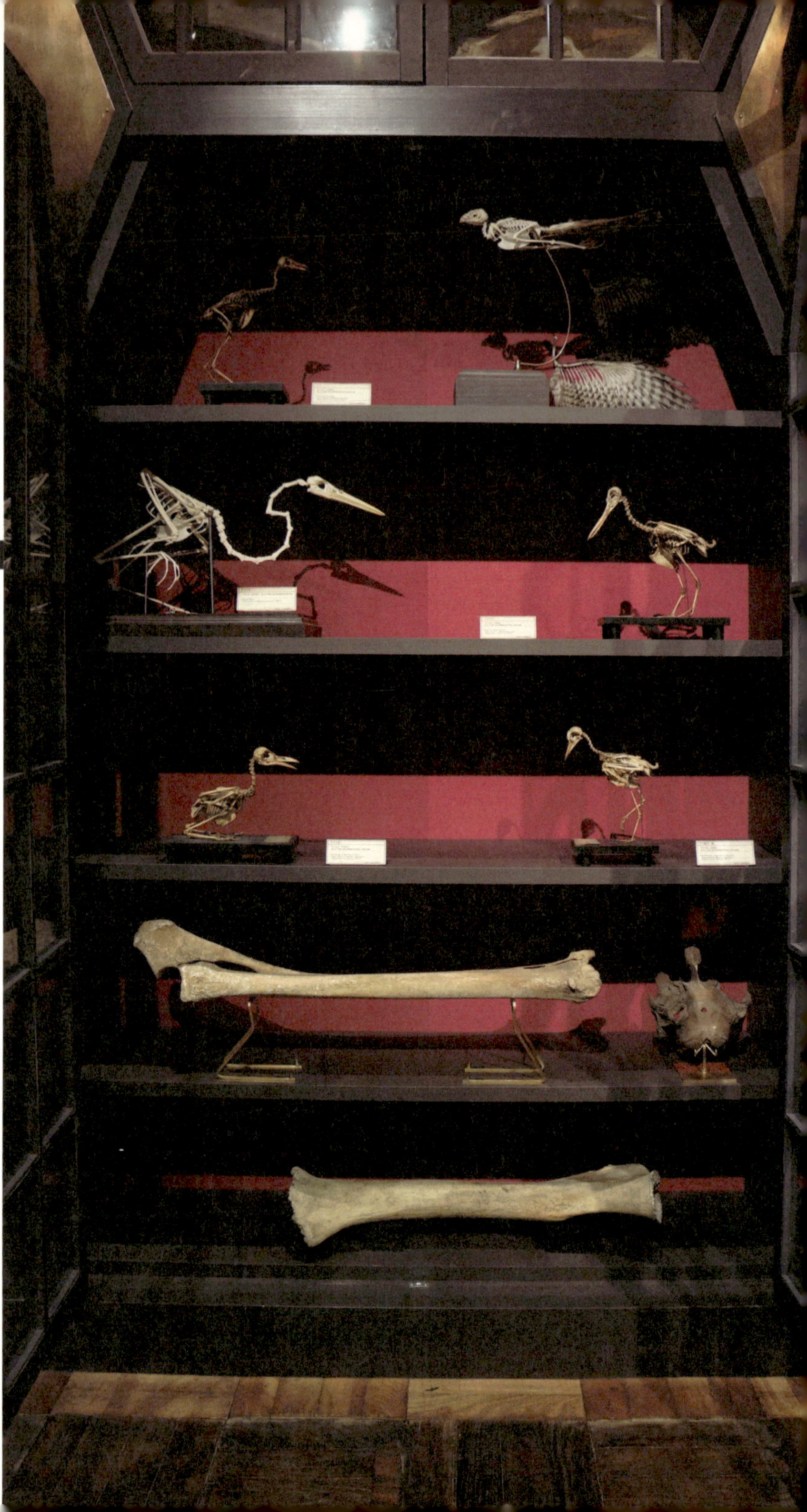

ソリハシセイタカシギ
Pied Avocet
Date unknown / Skeleton specimen /
Department of Medicine, UMUT

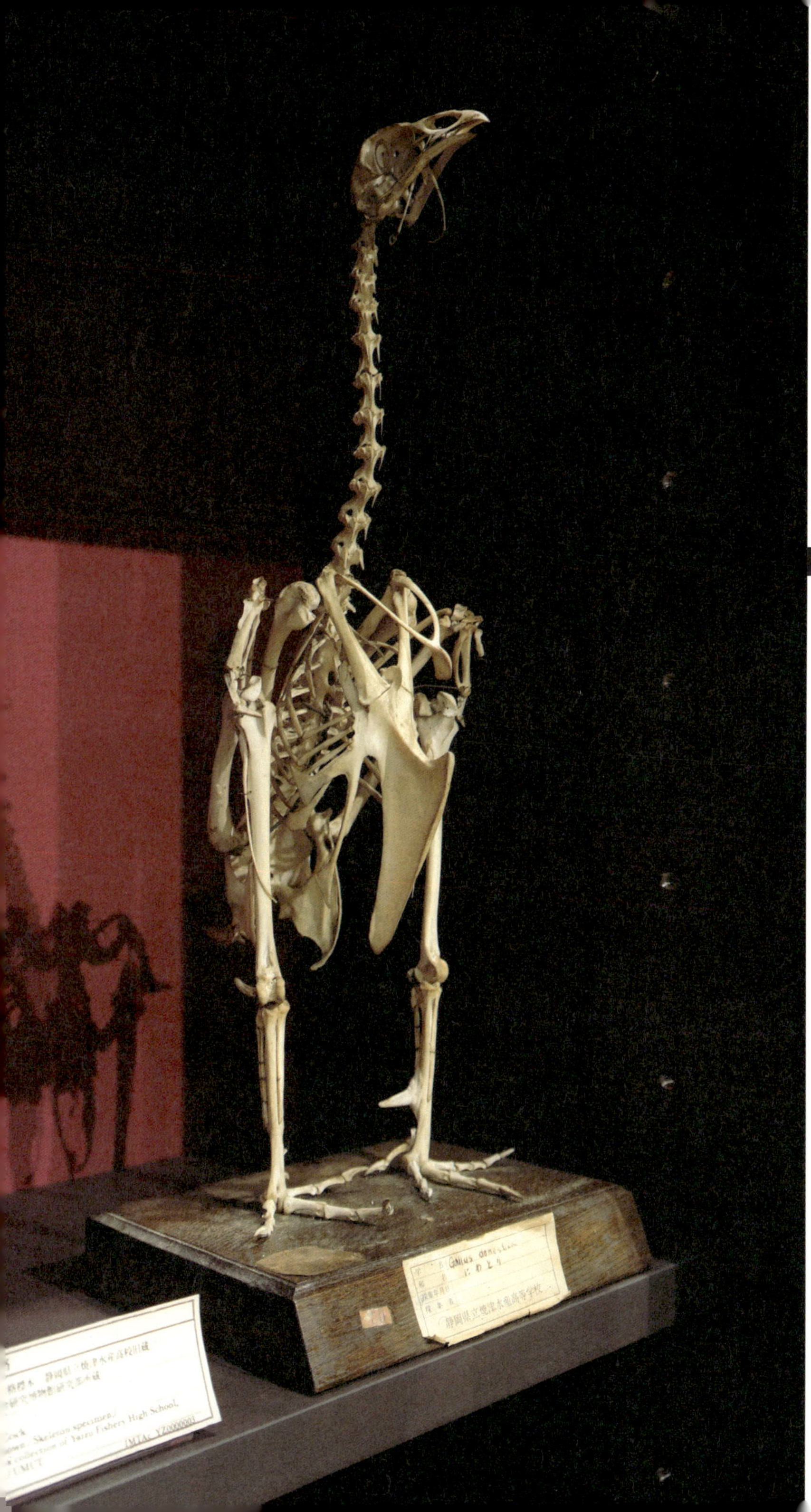
にわとり
静岡県立焼津水産高等学校
Skeleton specimen
Yaizu Fishery High School

Skeleton specimen/
UMUT

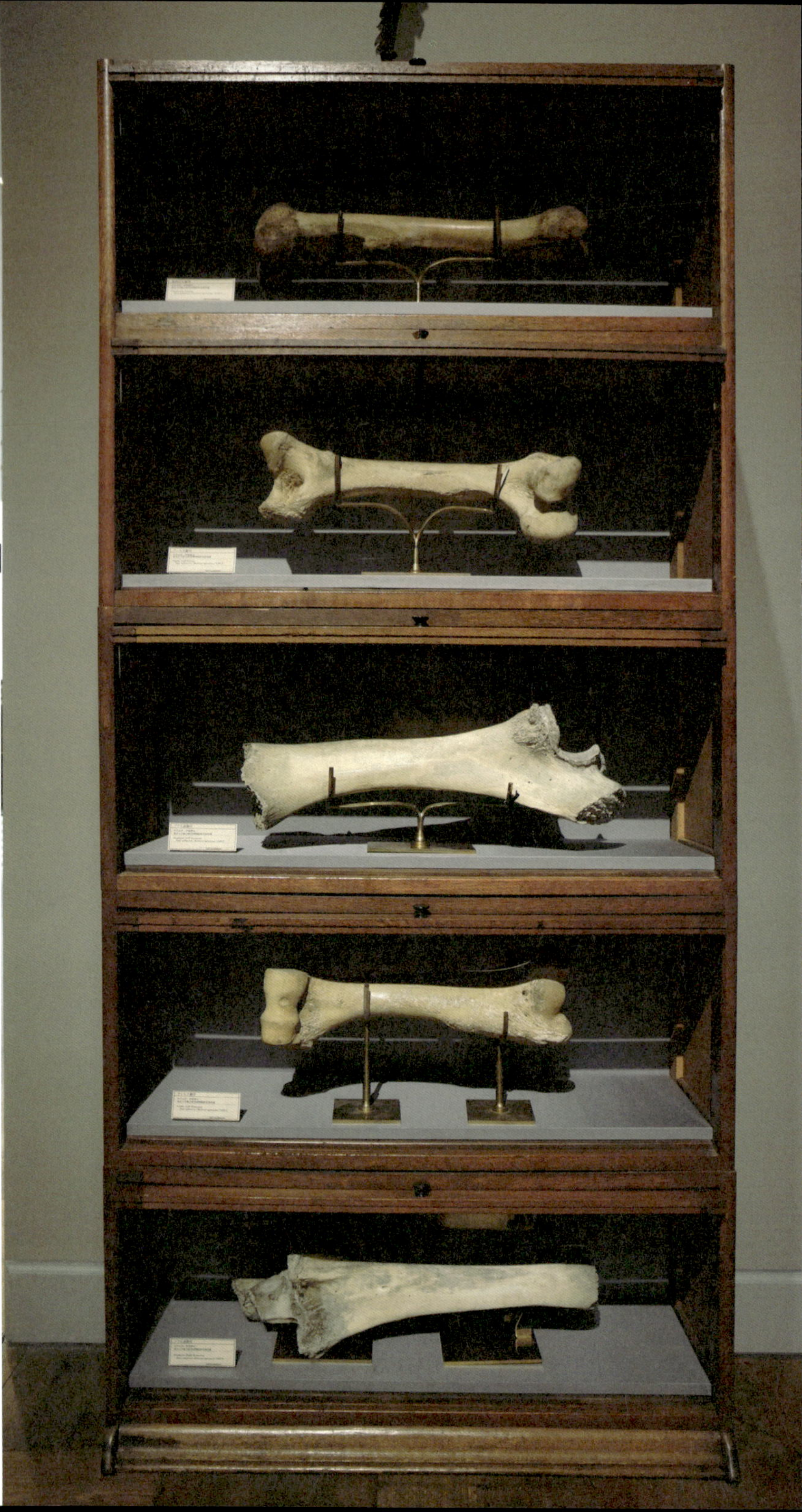

304

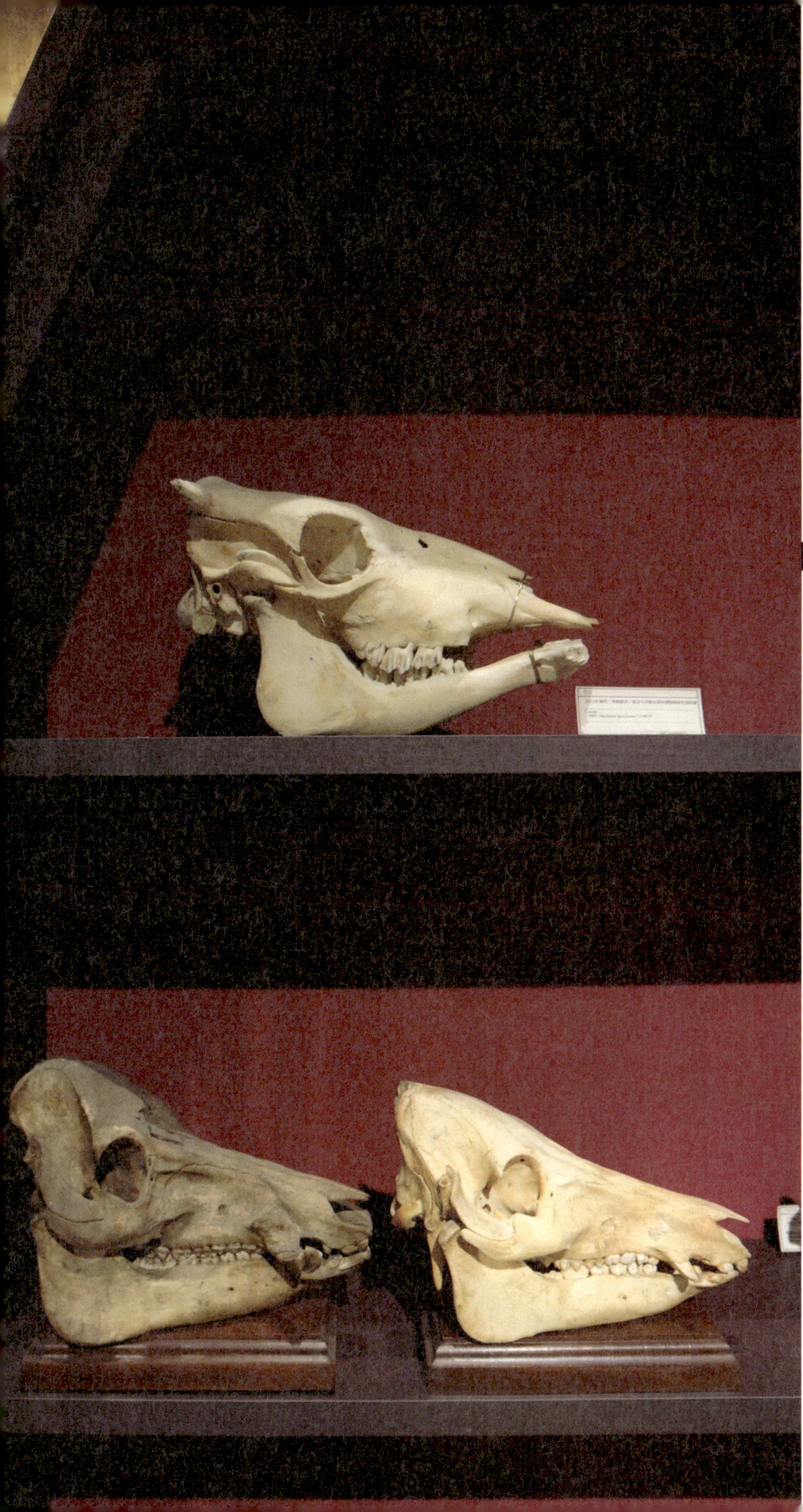

364
ヒツジ
年代未詳／骨格標本／
東京大学総合研究博物館資料部医学部門所蔵
Sheep
Date unknown / Skeleton specimen /
Department of Medicine, UMUT

Skeleton specimen / UMUT
モウコガゼル（雄）
Mongolian Gazelle, Male

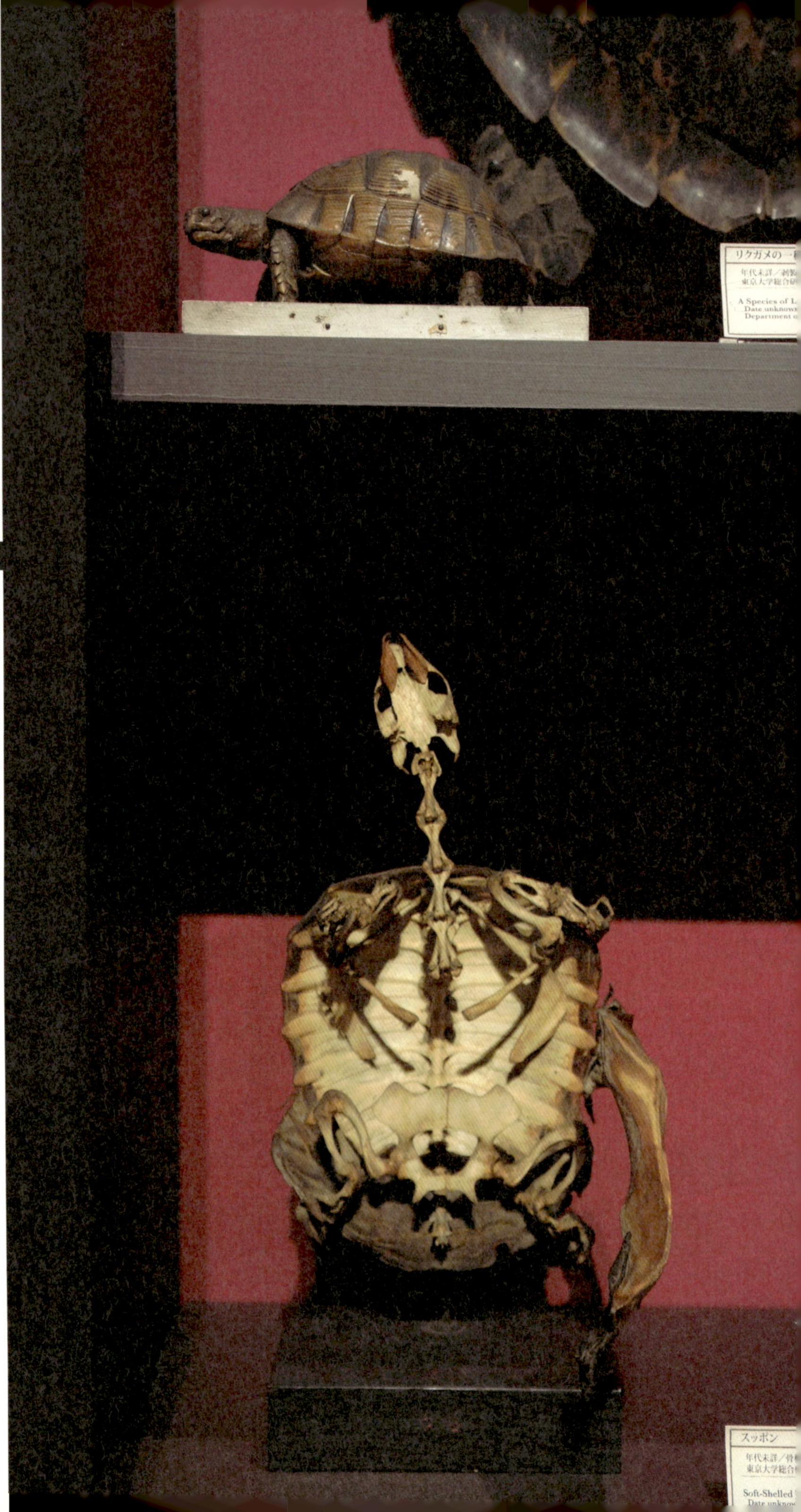
リクガメの一
年代未詳／剥
東京大学総合
A Species of
Date unknown
Department
スッポン
年代未詳／
東京大学総合
Soft-Shelled

ニホンイシガメ
年代未詳／骨格標本／
東京大学総合研究博物館資料部医学部門所蔵
Japanese Pond Turtle
Date unknown / Skeleton specimen /
Department of Medicine, UMUT
IMTE MD0000042

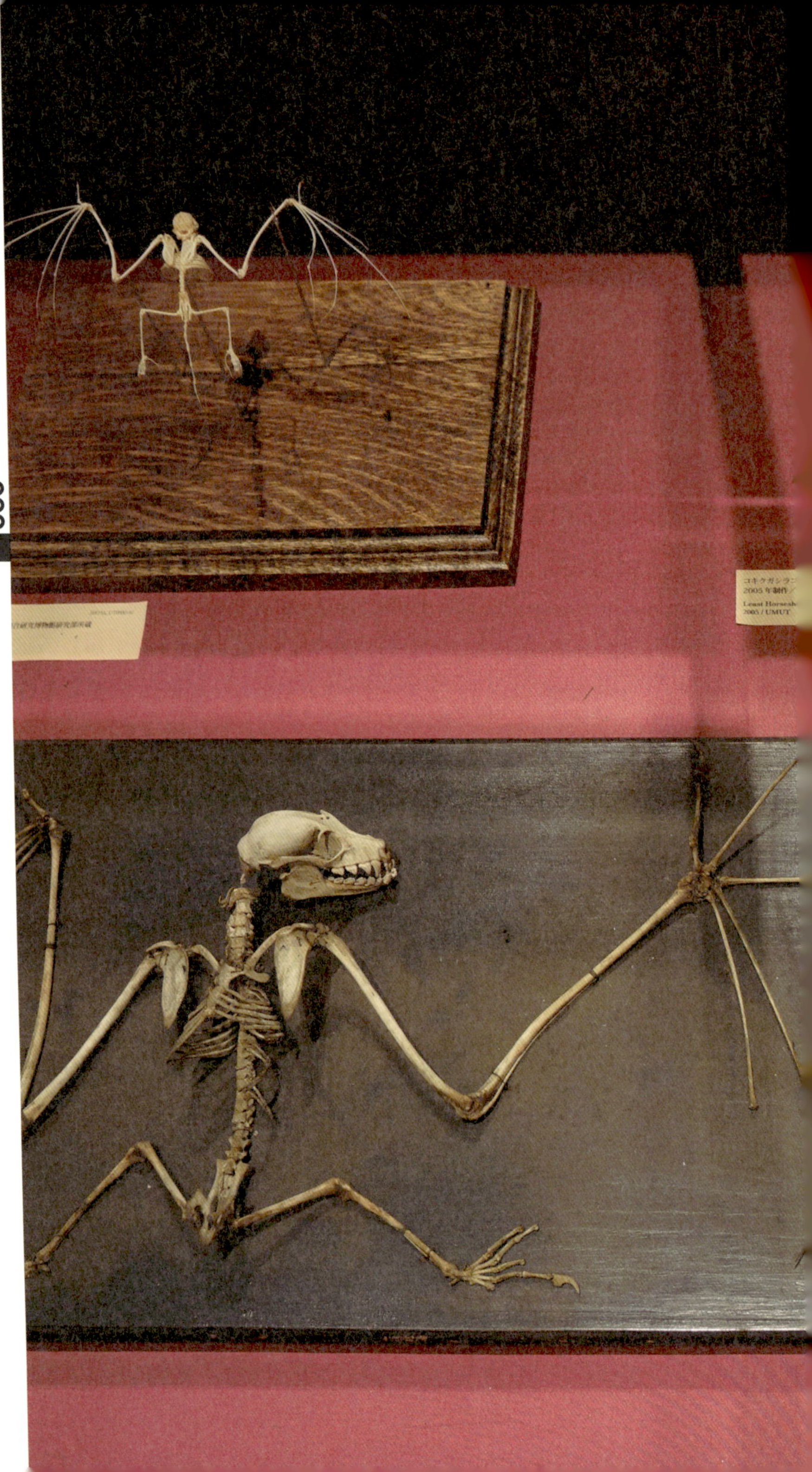

コウモリの一種
1968年／剥製標本／
東京大学総合研究博物館研究部所蔵
A species of Bat
1968 / Stuffed specimen / UMUT
IMTAb UT0000166
オオコウモリの一種
年代未詳／東京大学総合研究博物館資料部医学部門所蔵
Pteropodidae, a Species of Megabats
Date unknown / Department of Medicine, UMUT

ヨーロッパモグラ
年代未詳／東京大学総合研究博物館資料部医学部門所蔵
European Mole
Date unknown / Department of Medicine, UMUT
ツバメ
年代未詳
Barn Swallow
Date unknown
アカネズミ（雄）
2005年制作／宮城県金華山島
Large Japanese Field Mouse
2005 / Kinkazan Island, Miyagi
ハタネズミ
2005年制作
Japanese Grass Vole
2005

モモンガ
2006年制作
Japanese Lesser Flying Squirrel
2006
東京大学総合研究博物館
小型動物骨格標
東京大学総合研究
Small Animal
UMUT
The University

Japanese Toad

アズマモグラ
Small Japanese Mole

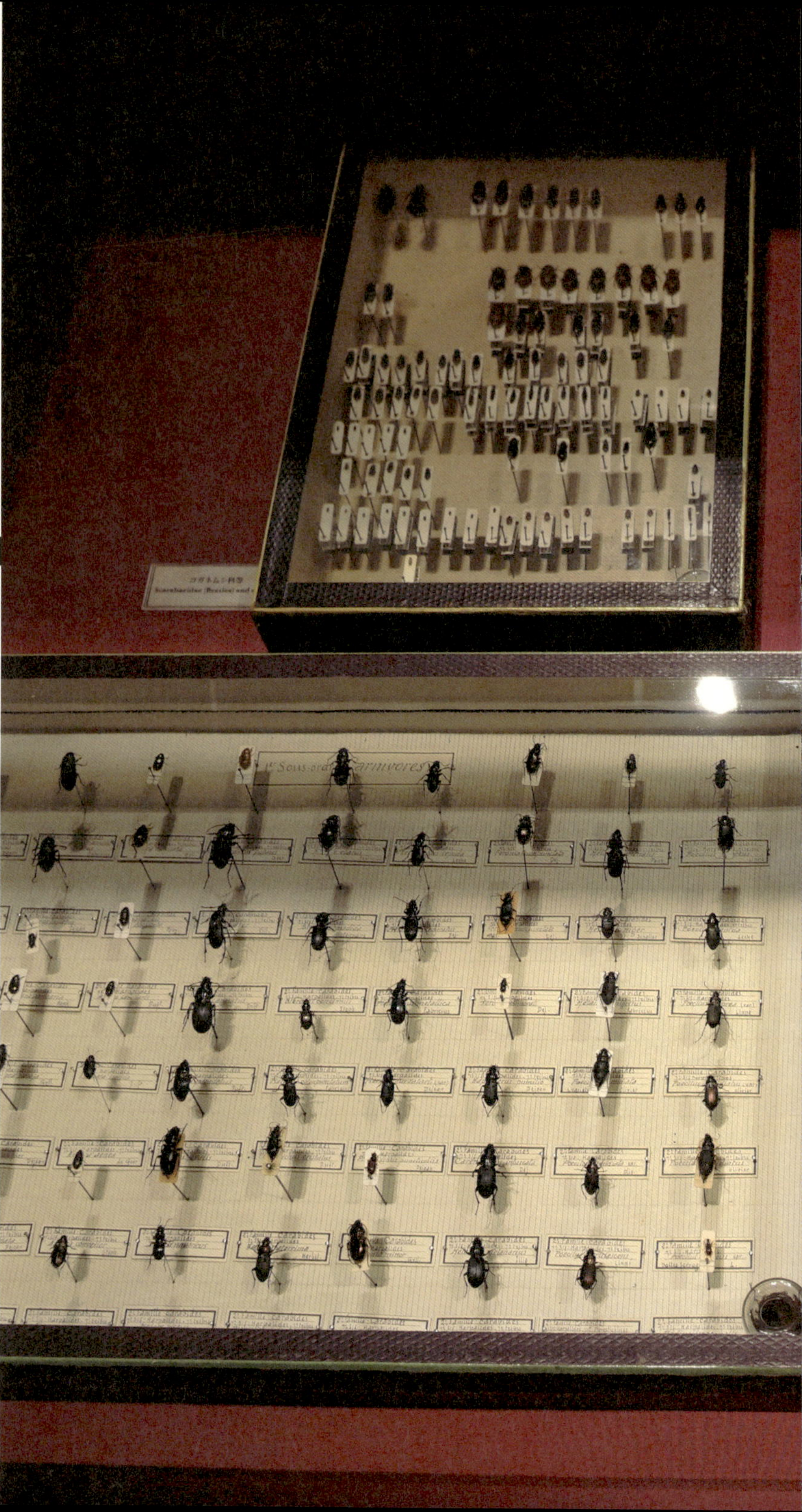
1er Sous-ordre Carnivores

東京大学総合研究博物館
フランス製甲虫標本コレクション
19世紀末から1970年代／2010年購入／乾燥標本／
東京大学総合研究博物館研究部所蔵
Beetle Specimen Collection from France
Late 19th Century to 1970's / Purchased in 2010 /
Dried specimens / UMUT
The University Museum - The University of Tokyo
肉食性甲虫
Carnivorous Coleoptera

カミキリムシ科
Longhorn Beetles

UNIV. MUS., UNIV. of TOKYO
貝殻標本コレクション
Representative specimens of molluscan shells collected by the Zoological Institute of the former Tokyo Imperial University. The institute started collecting various animals in the 1870's, the shell collection constituting the largest part. Specimens from all regions of the world were obtained through exchange with foreign institutes. Numerous tropical shells were also collected from Pacific islands by university staff. Conservation cases are made of paper-thin sheet of wood, with Japanese paper covering the wood and Japanese lacquer painted on the surface. This type of cases was most popular in Tokyo Imperial University.

Univ. Mus., Univ. of Tokyo
貝殻標本コレクション
20世紀後半／東京大学総合研究博物館研究部所蔵
Shell Specimen Collection
Late 20th Century / UMUT
貝類は昆虫に次いで種数が多く繁栄した動物群である。その種数は日本産だけで8千種を越え、全世界では正確な数値は知られていない。生息環境は主に海で波打ち際から深海まで幅広く生息し、さらに、汽水、淡水、陸上にまで進出している。東京大学では大学設立時より貝類の研究と収集を始め、それは現在まで途絶えることなく続いている。総合研究博物館は現在10万点以上の貝類標本を所蔵する。
Shells (molluscs) are the most specifically diverse animals after insects. The number of known species exceeds 8000 even within Japan, and little is known about the exact number of species in the world. Their habitats are mainly in the sea from the intertidal zone to the deep sea, but they also live in blackish and fresh water as well as on land. The University of Tokyo has conducted research and collection of shells ever since its foundation. The University Museum now preserves more than 100,000 shell specimens.

コジャコウネコ
Small Indian Civet

タイワンヤマネコ
Chinese Leopard Cat

高山市石浦地内
ニホンカワネズミ

ニホンリス
Japanese Squirrel
Date unknown /
Formerly Oita Wild Bird Museum Collection /
856

ニホンイタチ(雄)
Japanese Weasel (Male)

リス幼獣
年代未詳／（旧）老田野鳥館旧蔵／
東京大学総合研究博物館研究部所蔵
Japanese Squirrel (Juvenile)
Date unknown /
Formerly Oita Wild Bird Museum

オコジョ
年代未詳／（旧）老田野鳥館旧蔵／
東京大学総合研究博物館研究部所蔵
Stoat
Date unknown /
Formerly Oita Wild Bird Museum Collection / UMUT

ハリネズミ
年代未詳／東京大学総合研究博物館研究部所蔵
Hedgehog
Date unknown / UMUT

モモンガ
年代未詳／（旧）老田野鳥館旧蔵／
東京大学総合研究博物館研究部所蔵
Japanese Dwarf Flying Squirrel
Date unknown /
Formerly Oita Wild Bird Museum Collection / UMUT
IMTA60000139

IJIMA
MEMORIAL
LIBRARY

CRUSTACEA ARACHNOIDS MYRIOPODA
GENERAL AND MISCELLANEA

ゴホウラ
年代未詳／産地未詳／乾燥標本／
東京大学総合研究博物館研究部所蔵
trombus latissimus
Date unknown / Locality unknown / Dried spe
UMUT

テングニシ
年代未詳／日本／乾燥標本／
東京大学総合研究博物館研究部所蔵
Hemifusus tuba
Date unknown / Japan / Dried specimen / UMUT
シャゴウ
年代未詳／沖縄県西表島／乾燥標本／
東京大学総合研究博物館研究部所蔵
Hippopus hippopus
Date unknown / Iriomote Island, Okinawa /
Dried specimen / UMUT

アカニシ
Date unknown / Japan / Dried specimen / UMUT
テングガイ
UMUT
ヒレジャコガイ
年代未詳／産地未詳／乾燥標本／
Tridacna squamosa
Date unknown / Locality unknown / Dried specimen /
UMUT

Univ. Mus., Univ. of Tokyo
ウニ類標本コレクション

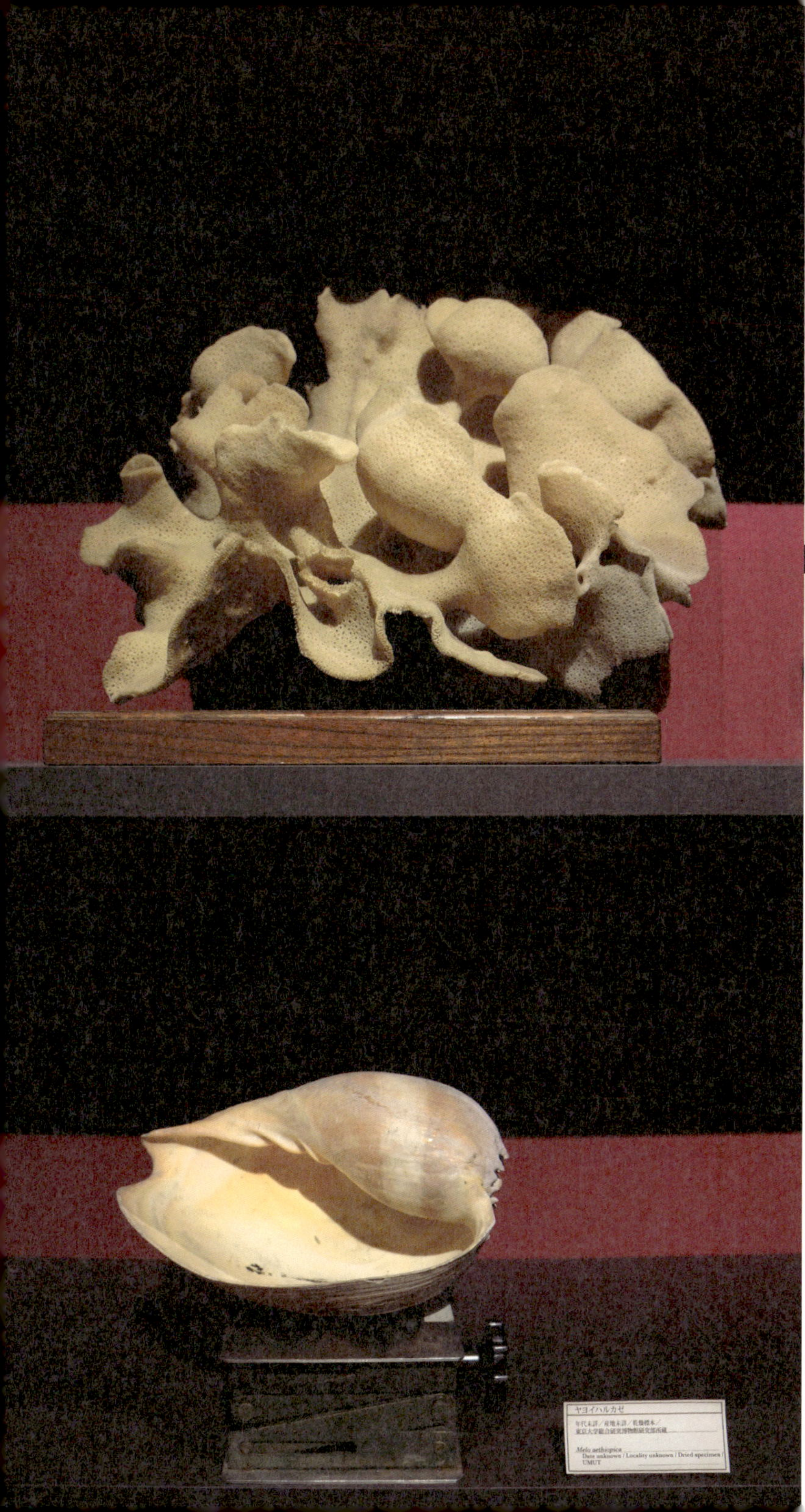
ヤヨイハルカゼ
年代未詳／産地未詳／乾燥標本／
東京大学総合研究博物館研究部所蔵
Melo aethiopica
Date unknown / Locality unknown / Dried specimen /
UMUT

urchased in 2011 / Philippines / Dried specimen /
epartment of Research, UMUT
ウニ類は棘皮動物門ウニ綱に属する動物である。体は球型、楕円形または円盤状の殻をもつ。生時は殻の表面が様々な長さの棘に覆われているが、死後は脱落し殻が剥き出しになる。ウニ類は深海から浅海のあらゆる海洋環境に適応しており、日本に約160種、世界に約900種が知られる。岩礁に生息するグループと堆積物中に生息するグループがあり、前者は海藻、後者は堆積物中の有機物を餌とする。
Sea urchins are
Echinoidea in the phylum
a calcareous test which is
test is covered with spines
animal's death. Sea urchins
habitats from shallow to
around 160 in Japan and
graze on algae, and those
particulate organic matter.
UNIV. MUS., UNIV. of TOKYO
ウニ類標本コレクション
Sea Urchin Specimen Collection

サンゴは刺胞動物の一種であり、イソギンチャクやクラゲに近縁である。動物
体は炭酸カルシウムの骨格を分泌し、サンゴ礁を形成する。大多数のサンゴ
は、遺伝的に同一な小さな個体（ポリプ）が集まってできる群体である。サン
ゴ自体はプランクトン食であるが、体内に褐虫藻を共生させており、褐虫藻
が光合成により生産する有機物を利用して成長することができる。そのため
には十分な光が必要であり、サンゴの生息域は熱帯域から亜熱帯域の浅海
に限られている。
Corals are a group of cnidarian animals which are phylogenetically close to sea anemones and jelly fish. Living animals secrete a hard skeleton of calcium carbontate and form coral reefs. The majority of corals consist of genetically identical, small individuals called polyps. Each polyp feeds on planktons and also uses organic nutrients produced by photosynthetic symbiotic microalgae (zooxanthellae). Because the algae need intensive sunlight, the distribution of corals is restricted to shallow-water tropical to subtropical areas.

UNIV. MUS., UNIV. of TOKYO
サンゴ標本コレクション
Coral Specimen Collection

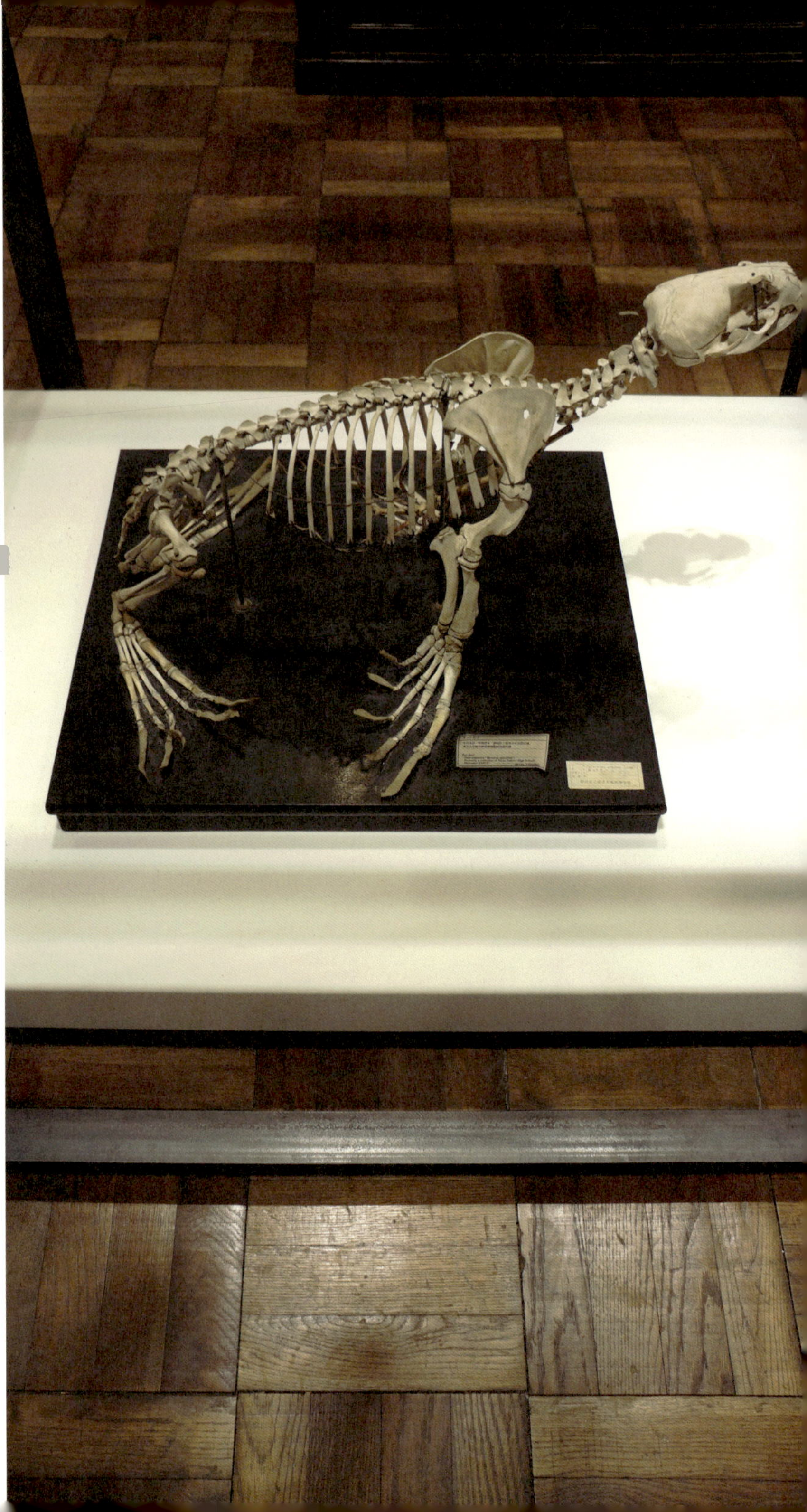

輝安鉱
Stibnite

N O R D

A 201
Collected by S.S. Haldeman
U.S.A
A 299
U.S.V.A
New Jersey

インターメディアテク

INTERMEDIATHEQUE

ごあいさつ

インターメディアテクは、日本郵政グループと東京大学総合研究博物館の協働運営になる公共施設であり、学術の普及と啓蒙を通じ、広く社会へ貢献することを使命としています。

このミュージアムには、東京大学が明治10（1877）年の創学以来蓄積を重ねてきた学術文化財が常設されています。展示に用いられているケースやキャビネットは、大方が研究教育の現場で使われていたものです。帝大時代の遺産が多く、それらのかもし出す重厚な雰囲気に、19世紀へタイムスリップしたような気分に捉われる方もあろうかと思います。ですが、われわれの企図するところは、決して懐旧的なものではありません。近代産業社会の誕生した19世紀から高度情報社会を実現した21世紀まで、その3世紀にまたがる時代を、あるいはサイエンスにおいて、あるいは創造的営為において、架橋することにあるからです。来るべき時代の精神がこの先なお見失ってはならない「世界の眺望」、それを東京丸の内という大都会の真中で開陳してみせることが狙いなのです。

ミュージアムというのは、これまで文化財の保管庫であり、収蔵品の陳列場であると考えられてきました。しかし、そうした機能を充足していただけでは、21世紀という時代の要請に応えることができなくなっています。現代のミュージアムは、われわれ人間が自分を取り巻く世界をどのように受け止めているのか、その俯瞰的な眺めを末永く存続させるための場所でなくてはなりません。それと同時に、その「俯瞰的な眺め」の構成要素として一堂に集められたモノやコレクションから、どのようにして新しい発見、優れた叡智を導き出すことができるか、その可能性を探求し、具体的なかたちをもって例証してみせる場所でもなくてはならないのです。学術研究の実践を通じて、多様な表現メディア間の対話を促し、新たな成果の誕生を誘う実験場、それがインターメディアテクなのです。

大学に蓄積されている学術標本は、たしかに、過ぎ去った時代の遺産かもしれません。しかし、同時にまた、われわれが未来に向けて活用すべきリソースでもあります。そのことを実証してみせたいと考えたわれわれは、歴史的な遺産を可能な限り収集蓄積し、それらを現代のニーズに適うよう、装いを改めて再利用することにしました。資源獲得やエネルギー供給に限界が見え始めた今日、蓄積財の「リデザイン」による活用可能性の探求は、喫緊の人類史的課題であるといって過言でありません。こうした現状認識に立って、われわれは先端的なテクノロジーと伝統的なもの作り技術の融合を図りつつ、「学術標本」を機軸とする各種の営為と取り組んでいくことになります。われわれの謳う「東京大学総合研究博物館自製」（Made in UMUT）の掛け声には、来るべき世代に向けてのささやかなメッセージが込められているのです。

インターメディアテクが、多くの方々に愛され、その支援にささえられつつ、発展しますことを念じてやみません。

館長
西野嘉章

「学術標本」と展示デザイン

西野嘉章

インターメディアテクの空間に布置されているモノは、一般のミュージアムで見られる展示物となにか違うのではないか、そのような印象をもたれた方もおられるかと思います。それもそのはずです。会場を歩いていただくとすぐにおわかりになろうかと思いますが、絵画にしても、標本にしても、いわゆる名物とみなされるようなモノはどこにも見当たりません。そればかりか、展示の仕方もふだん眼にするものと違って見えるはずです。年代に準じたり、種別に従ったりという、分明なやり方を採っていませんし、系統立てて並べようという意図もまた、端からもち合わせていないからです。

たしかに、インターメディアテクの展示には戸惑いを感じる方がおられるかもしれません。観覧の順路が明示されていないばかりか、どれが陳列物で、どれがそうでないのか、渾然一体として判別しづらいという声も聞こえてきそうです。いずれにせよ、インターメディアテクの展示が常套的なやり方に倣っていないことは確かです。どうしてなのか。この理由の一端は、インターメディアテクが一般のミュージアムと、成立事情も違えば、社会的な使命も異なる、大学の共同利用研究施設としての博物館、すなわち「大学博物館」（ユニヴァーシティ・ミュージアム）だということにあります。もっとも、「大学博物館」といっても、その実像をすぐに思い浮かべられる方は少ないと思います。平成 8（1996）年春、東京大学に発足した総合研究博物館は、先端諸学を推進する研究教育機関であると同時に、「デジタルミュージアム」「モバイルミュージアム」「学校モバイル」など、新しい学芸事業のあり方について、各種の実践を通じて社会提言をおこなう先導的な実験ミュージアムである、そのようにここではひとまずご理解下さい。

そうした前件がありますから、インターメディアテクにこれまで通りのミュージアム概念を当てはめようとすると、訳のわからぬことだらけということにもなりかねません。しかし、ここで立ち止まって、考えてみて下さい。現在の日本には六千館近い博物館ないし博物館相当・類似施設が乱立しています。ここでは個性を売り物にしている群小の博物館、あるいは個別のテーマに特化されている私立の博物館については触れぬことにしますが、国公立館の多くで開催されている展覧会は、企画展に冠されるタイトルこそ違え、展示コンセプトの面でどれもよく似たものばかり、というのがミュージアムの現状なのではないでしょうか。それもそのはずです。ミュージアムはこうあるべき、ミュージアムの展示はこうでなくてはならない、といった類の硬直化したミュージアム観に、学芸員をはじめとするミュージアム関係者はもとより、また学芸活動の果実を賞味する公衆までもが縛られてしまっているからなのです。そこで、こう尋ねてみたいのです。展示場の設えはどこも似たようなものであって良いのでしょうか、それぞれの館で個性的な世界が創出できるよう、もっと知恵をしぼらなくて良いのでしょうか、来場者の側もまた、あらかじめ設えられた順路を決められた通りに歩くことで、持ち前の知的好奇心を満足させることができるのでしょうか。現代のミュージアムの展示にあってはわかり易いことが善とされているようですが、「わかること」が鑑賞体験の至上の見返りであるとするなら、その先になにが展望できるというのでしょうか。実験館を標榜するインターメディアテクの立場から、問い質したいことは山ほどあります。

ところで、展示場を訪れるという体験は、はたして頭による理解だけのものなのでしょうか。「頭による理解」とは、文字や言葉を介して、主知的に物事を把握し、理解しようとする受容態度のことです。ミュージアムにおける鑑賞体験には、そのような唯脳的な受容しかあり得ないのでしょうか。ここにインターメディアテクからの問いかけがあります。

展示されたモノとの出合い。それは、まず眼で対象を捉えることに始まり、つぎにそれがなにか、脳に蓄積された記憶と照らし合わせながら認識するという段取りを踏みます。視覚による対象把握というのは、多分に記憶喚起的であり、運動感覚的なものです。となら ば、もっと感性に働きかけるような、でき得るなら身体の全感覚を総動員させるような鑑賞体験へ誘う展示があっても良いのではないでしょうか。

ミュージアムはこうあるべきだと端から決めてかかると、展示や企画は一様のものとなり、すぐに陳腐化し、やがて飽きられてしまいます。本来なら知的好奇心を活性化させる場所としてあるべきミュージアムが、文化生態系のなかで劣化していく。そうした悲喜劇的な事態が公共文化施設の周辺で進行しているように思えてなりません。均質化した文化環境は、創造のための胎土としては貧しい、このことの再確認が必要なのではないでしょうか。思えば、自然界における生物は多様であり、その多様性に自然の豊かさが示されています。それと同様、ミュージアムも各館それぞれが展示方針の違いを競い合ってこそ、文化的に豊かであると言えるのではないでしょうか。文化の母胎として豊かさを誇ろうとするなら、なによりもまず、他との差異を明示できなければならないと思うのですが、どうでしょうか。大学博物館は大学の研究教育機関であり、パイロット・ミュージアムです。このため、他に追随することもなければ、旧套的な手法を無反省に繰り返すこともしません。それこそが自分たち独自の方法にこだわり、その模索を続けてきた理由なのです。

モノの集結

大学では教育研究のかたわら、多くのモノが集められ、生み出されてきました。もちろん、こうした収集と生産の営みは、今もなお不断に続けられています。ところで、ひと口にモノを収集すると言いますが、大学の場合はどのようなかたちをとってきたのでしょうか。実際のところ、大学における収集様態は様々です。しばらく前までは、今ほど環境保全や文化財流出に関する取り決めが厳しくありませんでした。そのため、海外に出た研究者は、世界各地から研究資料を比較的自由に持ち帰ることができました。個人レベルの調査旅行の場合はともかく、大学が組織した大規模な学術調査隊の持ち帰り資料は、量もまた尋常でありません。百単位、千単位で毎年のように採集の続けられている植物標本のような場合、収蔵品も増加の一途を辿るばかりで、帰国後、十年、二十年と記載分類作業が続けられることも、決して珍しくなかったのです。もちろん、事は植物分野だけに限りません。動物学や鉱物学や地質学、あるいは考古学や人類学や民族学でも同じです。有力な資料体がかたちをなすまでには、そうした長期にわたる地道な作業の継続が必要とされていたのです。

大学では、必要に応じて、研究資料を購入する事例も少なくありません。古い話になりますが、維新後の明治新政府は近代国家建設の先兵となる人材を育成しようと、学制を整えるより先に欧米先進国から大量の標本や模型を買い付けました。実際のモノを前にして教育をおこなう。この瞠目すべき教育効果は、昔も今も変わりません。明治新政府はそのための最新の教育教材をいち早く揃えようとしたのです。その教育政策はまことに理に適ったものでした。当時の学術官僚の判断は実に賢明なものだった、と今にして思います。購入標本は、もちろん、研究に役立てられることもありましたが、なによりまず教育用の教材だったのです。実物を、あるいはサンプルや、模型や、複製をまず見せる。モノを見せることによって、より確かな理解を促す。こうした教育課程で用いられるモノは、一般に「参照標本」と呼ばれます。教材として学生に見せることを前提にして収集、製作、購入されたものですから、「参照標本」には典型的なモノがひと通り揃っています。系列的に整備されているばかりか、すがたかたちも申し分ありません。ですが、今やどうでしょう、実物教育が眼に見えて衰退し、文字情報とデジタル画像による擬制教育が主流になっています。伝統的な学術ツールとしての「参照標本」は活用される機会がなくなり、研究室や教室でお荷物扱いされる始末です。廃棄されず、総合研究博物館に回収ないし管理換されたモノがあるのは、不幸中の幸いと言うべきかもしれません。それら救済された「学術標本」が、インターメディアテクで陳列に供され、今ふたたび、教材としての真価を、そしてなによりもモノとしての魅力を発揮し始めているのです。

また、学外から、大学に対し、活用や公開を目的として、モノやコレクションがまとめて寄贈・寄託されることもありました。当然のことながら、恩恵者のなかには、卒業生や教員などの大学関係者、あるいはその御遺族の方々が多かったわけです。こうしたケースでは、モノの来歴や由来など、出自が確かかどうかが学術的な価値を左右すると言われます。とはいえ、学術的な価値というのは、あくまでモノの価値全体の一部に過ぎません。ですから、拙速な価値判断は禁物です。後になって、それまで知られずにあった価値が発見されることもあります。また、五十年、百年と時間が経つうちに価値を増すモノもあるに違いありません。ある専門分野でゴミとして処分されたモノのなかに、別の専門分野の研究者が、素晴らしい価値を再発見したというケースもあります。大学博物館が可能な限り多くの寄贈・寄託を受け入れるようにしているのは、そうした先例の数多あったことを教訓として学んでいるからです。モノが、既定の価値をアプリオリに内在しているわけではありません。モノの価値を決めるのは、他でもない人間の側なのです。われわれがモノの裡にそれまで知られていなかった価値を見出したとき、モノはそこで新たな価値を付与されることになるです。

モノの生産

モノが生産される様態もまた、変化に富んでいます。大学では論文を書いたり、研究を発表したりする活動が日々続いており、そうした営みの後には必ずモノが成果物として残されました。講義や実習をおこなうことも、当然、モノの生産に結びつくわけです。こうした研究現場では実験や試行をおこなうための、実験器具や補助道具が欠かせません。作図・計測・計量のための器具も必要となるはずです。ここでもまた明治初めの事例を引きますが、工部大学校の前身である工学寮の遺産のなかで現存最古のモノが、作図や計測のための道具類であるというのは、はなはだ意味深いものと言えます。国内における近代化が学術研究の基礎固めから始まったことを物語っているからです。日本では近代国家百年の計が、「モノサシ」の調達から始まったのです。

研究はモノに始まり、モノに終わると言われます。事の真偽はともかく、あのニュートンが林檎の落下を見て万有引力の発見に至ったというエピソードが語っているように、新しいアイデアの誕生の背後には、つねにモノの存在が前件としてあったのです。ひとたびそのアイデアが具現されると、事の道理として、また別なモノが生産されるわけです。理工や医学の研究室では、毎日のように試料や試作品や模型が生み出されています。直接的な研究対象を一次資料と言いますが、教育と研究に関わるのは一次資料だけに限りません。研究の結果として二次資料が生まれ、さらにはそこから後続の資料が派生します。このように研究や教育の現場では、連鎖的に資料がいや増しに膨らんでいくのです。

大学にモノが蓄積される理由は、それればかりではありません。教室、実験室、研究室など、教育研究の現場では、最低限の備品として机や椅子や什器が必要ですし、各種の研究資料や文書類を架蔵するためのキャビネットや書棚もなくてはならないのです。そうした家具類は、調達時に番号がふられ、学内のそれぞれの部局で、備品として登録されてきました。ですから、東京大学のような官制大学の調達品には、部局名、登録番号、登録年月日の記された小片が附されていたのです。幸いにして、そうしたものが残されている場合には、モノの出自を跡づけることができます。しかし、だからといって、そうした登録備品が後々まで生き残ったかというと、そうでもありません。調達費の原価償却が済むと、備品本体から出自を示す小片が外され、廃棄されるのが通例だったからです。もちろん、すべてが機械的に処分されてきたわけではありません。廃棄するのはもったいない。そうした当たり前の感情は、どの時代にも、誰の心にもあったからです。とはいえ、備品のリニューアルという口実のもと、貴重なモノが廃棄処分されてしまった事例はいくらもあったのです。

モノとしては、大学の運営全般に関わる公文書、記念アルバムや教授の肖像、胸像、さらには記念品、徽章、制服、顕彰碑など、大学の公史、教官の個人史に関わるモノや記録の類もあります。「大学史史料」と呼ばれるものがそれです。元帝国大学総長加藤弘之の個人史料をはじめとする、歴代総長の功績に関わる記録もまた、このカテゴリーに属します。こうした、ひとまとまりの資料体には、文庫、コレクション、史料などの冠名が附されます。分散させることなく、一ヶ所に集めて架蔵するわけです。第十四代総長内田祥三が関東大震災での被災を機に撮り始めた構内建築写真コレクション、あるいは大学本部に残されている学内施設建物の図面類といったものも、大学の存在、施設建物の変遷、先達の事績に深く関わる根本資料と言えます。

モノの活用

このように大学がその教育と研究の歩みのなかで生み出してきたモノは実に多様です。大学博物館では、これらをすべてまとめて「学術標本」と呼び慣わしています。「学術標本」は、当然のことながら、組成も寸法も形状も様々です。ですが、多岐にわたる教育研究活動の中身を反映しているという点はどれも共通しています。そればかりか、寄贈品や調達品は別にして、「学術標本」の多くは教員や研究者が自らの手で収集し形成した資料ですから、出自来歴が確かであり、科学的な検証にも耐え得るという特徴があります。この点が、出自来歴について第三者の記述に頼らざるを得ない、一般の博物館コレクションと性格を異にしているのです。後に残された「学術標本」をつぶさに検証すると、それを生成した研究者の仕事ぶりが見えてきます。未来を担う研究者は、先輩たちの残した「学術標本」を一覧することで、現在の自分の立ち位置を俯瞰的・通時的に把握し、確認できるわけです。

法科、工科、医科、理科、農科など、個別の専門領域に特化された単科大学の場合、「学術標本」の多くはその専門分野の研究や教官に関わる資料・記録が中心となります。総合大学の場合には、蓄積される「学術標本」もまた多分野に広がります。ですから、東京大学のように、国内でもっとも長い歴史を誇る大学の場合には、標本の分布域も蓄積量も、他に類を見ぬほどまで拡大するわけです。

ならば、実際のところ、東京大学にはどれほどの標本が蓄積されているのでしょうか。これについては、誰も正確な量を把握していないというのが、正直なところです。現に今もなお、不断に標本が生み出され続けています。いささか古い数字になりますが、旧総合研究資料館が現在の総合研究博物館へ改組拡充を果たした平成8（1996）年の統計によると、東京大学全体で六百万件が数えられ、そのうちの二百四十万件が館内に収蔵されていたことになります。これには、もちろん、一定の留保が必要です。「一件」という数字をどのように見るか、という問題があるからです。インターメディアテクの開館時、展示会場に並べられたモノを数えると七千点近くに上りました。件数でいえば千件程度と見積もられます。統計数値は必ずしも実態を表していないということです。また、上記の数字は登録の済んだモノの数ですから、未登録のモノを含めると全体数がどの程度まで膨らむのか、想像さえし難いわけです。附言するなら、総合研究博物館の場合、改組以来、年平均で五万件から十万件のペースで標本数が増えていますので、現状では優に四百万件を超えていると見積もられます。

インターメディアテクには、そうした膨大な蓄積量を誇る「東京大学コレクション」のなかから、稀少性や学術性において瞠目すべき「学術標本」を中心に選び、展示してあります。展示品は文化誌から自然誌、さらには工学や科学史など、多様な分野をカバーしています。時空間的な広がりにおいてもそうです。スターダストや隕石など、惑星科学のもたらす資料は悠久の宇宙的時間相に属しています。また、岩石学や鉱物学の標本は地球の成り立ちに関わる地質年代の尺度で測られるものです。先史人類学の化石人骨や石器は百万年から十万年の単位、古代文明の遺産は千年単位、美術・文化財は百年単位、いずれにしても人間的な時間の尺度で年代分布を把握できます。当然のことながら、空間的にも地球の全域をカバーする広がりを有しています。

「東京大学コレクション」は多様な分野にまたがるコレクションによって構成されています。ですから、コレクションの集合体（コレクション）と言うべきかもしれません。そればかりではありません。それぞれの集合体（コレクション）が、広く国内外の研究機関の同種のコレクションのなかにあって、第一級の位格を有するものばかりです。明治10(1877)年の創立以来、東京大学は官制大学として国内の教育研究を先導してきました。そのことを考えるなら、学内に蓄積されたコレクションが特段に優れたものであっても不思議ではありません。内国的な視点に立ってあえて極論するとすれば、学術研究の各分野において、最初の発見、最古の史料、最大の標本、第一号の試作物、稀少性・学術性において瞠目すべきモノ、それらの多くが「東京大学コレクション」のなかに集中的に存在していると言えます。それらの一部がインターメディアテクの空間を飾っているのです。

モノの開陳

大学博物館の収蔵品は、上記のような経緯のもとで一ヶ所に集められた「学術標本」のコレクションです。ですから、展示をおこなう段には、一般のミュージアムにしばしば見受けられるような名品至上主義を採るわけにいきません。とはいえ、名物品がないからといって挫けてなどいられません。名宝や名品に依存したくても依存できない。そうした前件を素直に受け止め、その上でなにができるか、自ら問わねばならなかったわけです。なにか新しいアイデアを導き出し、創意と工夫を施すこと。それが端から義務づけられているという現実を、これまでにない展示デザインを実現する上での生産的な動因と考えたらどうか。インターメディアテク構想は、そうした積極的な、建設的な考えに端を発しています。これはひとつの例に過ぎませんが、たとえば、自然誌展示の場合、異なる種類の標本を大量に並べてみせることで、種の多様性を示すことができますし、また同一種の標本を数多く並べてみせることで、個体の分布幅を示すこともできます。それ以上に、大学の教育と研究のプロセスのなかから生み出された「学術標本」には、どうしてこのようなモノに関心をもつに至ったのか、サイエンティストの発想の奇抜さや面白さの根源に触れる要素が溢れています。「学術標本」が素人眼にも興味深く映るとしたら、そうしたところに理由が潜んでいるのかもしれません。

ミュージアムにおけるモノとの出合いは、どうあるべきなのか。インターメディアテクの展示には、そのような問いかけが込められています。そもそもが研究と教育を目的に収集されたものですから、必ずしも鑑賞対象として相応しいものとは言えません。「学術標本」の多くは、個々物の質の高さより、コレクションとしての全体像に、見た目の面白さがあります。ですから、普通とは異なる見せ方を工夫しなくてはならないのです。

この「工夫」ということばのなかには、視覚心理学的な戦略、スペクタクル演出論的な効果、造形美術論的な創意、博物館工学的な構成など、多岐にわたる要素が含意されています。創意と独創性に溢れるものであること、そしてそれよりなにより鑑賞にたえるモノとして、見た目の美しさ、遭遇の心地よさをどのように担保したら良いのか、そうした問いかけの姿勢をもって、多角的な観点で課題解決を図りながら展示が構成されているのです。こうした意図のもと、インターメディアテクでは学内で長く使われてきた什器類、あるいは古い建物の部材が各所に活用されています。あるものは展示という用途に適うよう加工が施されています。また、建物を取り壊したときに出る廃材を使って作られた展示具もあります。こうした転用、加工、合成の操作を、インターメディアテクでは「リデザイン」と呼び、積極的に採り入れるようにしています。不要となったものをリサイクル活用することで、資源消費を抑制し、環境負荷を軽減しよう、それがインターメディアテクの基本的な考え方なのです。

すでにインターメディアテクをご覧いただいた方は、会場構成に順路らしきものが設定されていないことに気づかれたに違いありません。あらかじめ決められた順路に従うのでなく、自然誌の研究者がフィールドに出て動物や植物を探して回るように、インターメディアテクの来館者もまた、展示場のなかを興味の赴くまま、自らの足で自由に遊歩しながら、モノとの出合いを愉しむ。そうした場所になるよう、インターメディアテクは設えられているのです。

一般のミュージアムは社会教育施設と位置づけられています。そのため、展示品の選択や順序、解説文の中身や表示に、学習効果を上げようという配慮がなされます。ときには必要以上に。またあるときは無神経と言えるほどに。説明文をできるだけ懇切丁寧なものにしたり、大きな文字でそれを目立つように表示したり、というようにです。今日では、展示物のそばにデジタル端末を設置し、そこから附帯情報を引き出せるようにしたり、音声ガイドで順路に従って歩かせたりもします。あちこちでこの種の展示を見かけますから、たしかに現在のミュージアム界に普及しているのでしょう。しかし、学習教育、情報伝達という面でその種の展示が本当に効果的なのかどうか、検証がなされたという話を聞いたことがありません。ここでその妥当性について云々するつもりはありませんが、少なくともインターメディアテクでは、感性的な観覧環境を重視するという方法的な意識に則って、解説文等の文字情報ができるだけ視覚的な夾雑物とならぬよう、特段の注意を払っています。

デザインへの配慮の乏しい説明文や図説を、無神経に人前に晒して反省のない展示、そうしたものを憚りもなく公開することのどこが教育的なのでしょうか。国内のミュージアムにおける、あまりにもありふれた光景。インターメディアテクでは、そうした社会教育のあり方の「非文化性」を問いたいと考えています。知識の伝達を軽んじる気は毛頭ありません。ですが、審美性の乏しい展示デザインは百害あって一利なしです。ほんのわずか、デザインに心を砕くだけで良いのです。魅力的なモノを美しい姿で展示してこそ、感性の涵養に資するのではないでしょうか。インターメディアテクは読むことを通して概念を理解するミュージアムでなく、視ることを通して創造を惹起するミュージアムなのです。

005-006　マチカネワニ（IMTAb_UT0000129）
更新世／樹脂（骨格標本レプリカ）／原野農芸博物館寄贈／東京大学総合研究博物館研究部所蔵
1964（昭和39）年、大阪府待兼山の更新世（50万から30万年前）の地層から発見。

007　『空間の卵』のインスタレーション
2007年／西野嘉章＋セルジオ・カラトローニ＋関岡裕之＋中坪啓人／エピオルニス卵殻（京都科学複製）、大理石、真鍮、欅台座／東京大学総合研究博物館研究部所蔵
コンスタンティン・ブランクーシの『空間の鳥』の対位法的な解釈に基づくインスタレーション。ブランクーシは鳥をテーマとする一連の作品制作のなかで、卵のフォルムにも関心を払っていた。金属板の上に卵形フォルムを置いた作品も知られているが、卵を立たせた例はない。本インスタレーションを組み立てるにあたっては、台座等にできるだけ彫刻家の好個の素材を用いた。

008　電気工学器具（IMTF0000297）
年代未詳／金属／東京大学工学部電気工学教室旧蔵／東京大学総合研究博物館研究部所蔵

009　工学機器

010　クダマキガイ科の一種
1990年代／奄美大島（鹿児島県）／乾燥標本／東京大学総合研究博物館研究部所蔵

011　田口和美像（IMTE_MD0000198）
1894（明治27）年／高橋勝蔵画／布に油彩／東京大学医学部解剖学教室所蔵
東京大学内に現存する最古の肖像画。田口和美（1839-1904）は東京大学医学部解剖学教室の初代教授。1893（明治26）年に日本解剖学会が設立されると、その会頭を務めた。肖像画はこの頃のものである。作者の画家高橋勝蔵（1860-1917）は1885（明治18）年に渡米し、サンフランシスコのカリフォルニア・デザイン学校で油絵を学んだ。1893年に帰国し、東京芝に芝山研究所を開設した。

012　蒸気機関模型（IMTF0000036）
1870年代／エリオット・ブラザーズ社／ロンドン（英国）／東京大学総合研究博物館研究部所蔵
動力伝達の機構をわかりやすく理解させるための教育教材。

013　大熊氏廣　モーセ像（IMTJ_UT0000186）

1879（明治12）年／石膏／工部美術学校彫刻学科旧蔵／東京大学総合研究博物館研究部所蔵
大熊氏廣（1856-1924）は、工部美術学校の開校とともに入学し、ラグーザの下で彫刻を学んだ。在学中に制作した石膏像二点が現存する。そのうちの一点である本作品の背後には「明治十二年二月□□終　ミツシエラン□□　大熊氏廣」と記されており、ミケランジェロの『モーセ像』の模刻と思われる。

014　チャールズ・ディッキンソン・ウェスト像（IMTJ_UT0000185）
年代未詳／沼田一雅作／樹脂／東京大学工学系研究科産業機械工学専攻旧蔵／東京大学総合研究博物館研究部所蔵
チャールズ・ディッキンソン・ウェスト（1847-1908）は工部大学校雇用のアイルランド人教師。ダブリンに生まれ、ダブリン大学トリニティ・カレッジで機械工学を修め、1869年卒業。五年余り英国のベルゲンヘッド製鉄場で働いた。造船の知識をこの時に得ている。工部大学校都検ヘンリー・ダイヤーの後任として1882年に来日し、機械工学とともに造船学も教えた。そのまま日本に留まり、1908年に没した。

015　ウシの角
年代未詳／乾燥標本／個人蔵
ウシの角は骨質の角突起の上に角質の鞘が被さっており、この標本のように角鞘部分だけを取り外すことができる。このような角を洞角といい、生え変わらず成長し続ける。この点が内部まで一体化し、毎年生え変わるシカの枝角とは異なっている。角鞘は主にケラチンから成り、ヒトで言えば爪に似た構造である。標本は表面を磨いたもので、螺旋状に伸長した痕跡を読み取ることができる。

020-021　大型骨格標本

022　甲殻類乾燥標本コレクション
明治10年代／飯島魁か／東京大学（旧）理学部動物学教室旧蔵／東京大学総合研究博物館研究部所蔵

023　爬虫類・両生類剥製標本コレクション
明治10年代／佐々木忠治郎か／東京大学（旧）理学部動物学教室旧蔵／東京大学総合研究博物館研究部所蔵

024　スッポン（IMTAd_UT0000023）
年代未詳／剥製標本／東京大学総合研究博物館研究部所蔵

025　カエル交連骨格標本コレクション（IMTAd_UT0000001-13）
明治10年代／佐々木忠治郎か／東京大学（旧）理学部動物学教室旧蔵／東京大学総合研究博物館研究部所蔵
カエルの愛嬌のある姿の下には、跳躍のために特殊化した骨格が隠れている。特に腰から後肢にかけての骨格は他に例のない形態であり、強大な推進力を生む。一方、頭骨や体幹部の骨格は単純で原始的な形態である。

026-027　エピオルニス卵殻
来歴未詳／2012年複製／樹脂に彩色／仏国エクス=アン=プロヴァンス自然史博物館寄贈／東京大学総合研究博物館研究部所蔵
エピオルニスは、知られている動物種のなかで、最大の卵を産む種でもある。卵の容積は7から9リットルあり、鶏卵ならば約180個分に相当する。卵は自然界から生み出された完全無欠な「かたち」の代表であり、生命を裡に宿すことから、万物を内包した宇宙球にもたとえられる。

028-029　哺乳類骨格標本コレクション

030　ニホンイタチ（オス）（IMTAb_UT0000041）
2005年／骨格標本／東京大学総合研究博物館研究部所蔵

アナグマ（IMTAb_UT0000043）
2003年／骨格標本／東京大学総合研究博物館研究部所蔵

031　アカギツネ (IMTAb_UT0000001)
年代未詳／骨格標本／東京大学総合研究博物館研究部所蔵

アナグマ (IMTE_MD0000028)
年代未詳／骨格標本／東京大学総合研究博物館資料部医学部門旧蔵／東京大学総合研究博物館研究部所蔵

カワウソ亜科の一種 (IMTE_MD0000021)
年代未詳／骨格標本／東京大学総合研究博物館資料部医学部門旧蔵／東京大学総合研究博物館研究部所蔵

032　キョン (メス) (IMTAb_UT0000003)
1957年死亡／骨格標本／東京大学総合研究博物館研究部所蔵
ジャワマングース (オス) (IMTAb_UT0000040)
2001年／奄美大島／骨格標本／東京大学総合研究博物館研究部所蔵
タヌキ (IMTE_MD0000031)
年代未詳／骨格標本／東京大学総合研究博物館資料部医学部門旧蔵／東京大学総合研究博物館研究部所蔵

アナウサギ (IMTE_MD0000026)
年代未詳／骨格標本／東京大学総合研究博物館資料部医学部門旧蔵／東京大学総合研究博物館研究部所蔵

ハリネズミ (IMTE_MD0000027)
年代未詳／骨格標本／東京大学総合研究博物館資料部医学部門旧蔵／東京大学総合研究博物館研究部所蔵

033　ニホンジカ胎児 (IMTAb_UT0000057)
2003年／骨格標本／東京大学総合研究博物館研究部所蔵

034-035　哺乳類骨格コレクション

036　イエネコ (IMTAb_UT0000039)
年代未詳／骨格標本／東京大学総合研究博物館研究部所蔵

ケナガイタチ (IMTE_MD0000029)
年代未詳／骨格標本／東京大学総合研究博物館資料部医学部門旧蔵・東京大学総合研究博物館研究部所蔵

037　イタチ (IMTE_MD0000032)
年代未詳／骨格標本／東京大学総合研究博物館資料部医学部門旧蔵・東京大学総合研究博物館研究部所蔵

ヘビの一種 (IMTE_MD0000058)
年代未詳／骨格標本／東京大学総合研究博物館資料部医学部門旧蔵・東京大学総合研究博物館研究部所蔵

038　アカシカ (雄) (IMTAb_UT0000064)
2003年／モンゴル／骨格標本／東京大学総合研究博物館研究部制作・所蔵
ヨーロッパから中央アジア、アフリカの一部に分布する大型のシカ。雄は体高150センチ、体重200キログラム以上に達する。シカの特徴である枝角は骨質の構造で、秋の求愛シーズンが過ぎると毎年生え変わる。春になると新たな角が伸び始め、秋までには完成する。大きな角は雌へのアピールかつ雄同士のステータスシンボルであり、時には戦いに用いられる。本個体は総合研究博物館教員が海外調査先で入手したもの。

039　ダチョウ (IMTAb_UT0000150)
2005年 (推定) ／骨格標本／井の頭自然文化園旧蔵／東京大学総合研究博物館研究部制作・所蔵
現生の鳥類では最大。体重は130キログラム程度になる。足指は2本になっており、平地を走ることに特化した骨格である。飛ぶことはできないが、走行中のバランスをとるため、あるいはディスプレイとして翼を使うため、骨格は完全に退化しているわけではない。外敵から逃げるため、時速70キロメートルに達する快速で走ることもできる。

040 形象的形態（木彫人頭骨）
2011年／菊池敏正制作／檜／個人像
日本には古くから「木骨」の伝統がある。江戸時代末期まで、医学研究の現場で「真骨」を扱うことが、仏教上の制約もあり、禁じられていたからである。整骨医は、仏像制作を専門とする仏師に依頼し、「木骨」を用意し、それを教育研究に使っていた。現代の彫刻家が、その伝統の再興を試みたものである。

最初のラミダス化石
440万年前／1992年12月17日発見／エチオピア／研究資料（レプリカ）
ラミダス猿人（学名*Ardipithecus ramidus*）は、1992年当時、最古の人類化石として発見された。その後、2009年に全身像が発表され、人類最古の進化段階として知られるようになった。類人猿のように把握性の足をもちながら、人類のように直立二足歩行を行っていた。チンパンジーの系統と分岐後まもない状態を色濃く残していると考えられる。展示の標本は、エチオピアのアファール地溝帯のアラミスの調査で発見されたラミダスの第1号標本、右上顎第3大臼歯。

041 ペルー、クントゥル・ワシ遺跡の黄金墓の発見　紀元前約800年
クントゥル・ワシはペルー北部に位置するアンデス文明形成期中期から後期（紀元前約1000-250年）の神殿遺跡である。1988年より東京大学古代アンデス文明調査団が発掘調査を開始し、1989年から90年にかけて黄金製品を伴う墓4基を発見した。これらは精錬された黄金製品としてはアメリカ大陸最古のものである。展示品は3基の墓から出土した副葬品を、オリジナルと同じ材料組成により再現したレプリカである。
十四人面金冠／金製横顔ジャガー耳飾り／金製リング状耳飾り／五面ジャガー金冠／金製ジャガー・双子の鼻飾り／金製蛇目・角目ジャガー鼻飾り

042 バンドウイルカ（IMTE_MD0000035）
1957年9月／骨格標本／静岡県安良里／東京大学総合研究博物館資料部医学部門旧蔵・東京大学総合研究博物館研究部所蔵

043 スズキ（IMTE_MD0000011）
年代未詳／骨格標本／東京大学総合研究博物館資料部医学部門旧蔵・東京大学総合研究博物館研究部所蔵

044 ホタルジャコ科の一種（IMTE_MD0000019）
年代未詳／骨格標本／東京大学総合研究博物館資料部医学部門旧蔵・東京大学総合研究博物館研究部所蔵

045 フナ属の一種（IMTE_MD0000017）
年代未詳／骨格標本／東京大学総合研究博物館資料部医学部門旧蔵・東京大学総合研究博物館研究部所蔵

046 ホウボウ（IMTAe_UT0000002）
2012年／骨格標本／東京大学総合研究博物館研究部所蔵

047 ギンカガミ（IMTAe_UT0000003）
2012年／骨格標本／東京大学総合研究博物館研究部所蔵

048-049 ミンククジラ（IMTAb_UT0000126）
2009年／骨格標本／東京大学農学部旧蔵／東京大学総合研究博物館研究部制作・所蔵
体長7-8メートル、体重8トンほどになるが、ヒゲクジラとしては二番目に小型。鯨類は水中生活に適応する過程で後肢を退化させており、体重から解放された椎骨の形も単純である。鼻孔は巨大な頭骨の中央に移動し噴気孔となるなど、頭骨の特殊化も進んでいる。歯はなく、水とともに吸い込んだオキアミや小魚を上顎のヒゲと呼ばれる器官を用いて漉し取る。目はあまり発達しておらず、もっぱら音響や音声信号を用いて生活する。

050　ニホンザル (IMTAb_UT0000171)
2006年／骨格標本／東京大学総合研究博物館研究部所蔵

051　鳥類頭骨コレクション
年代未詳／東京大学総合研究博物館資料部医学部門旧蔵／東京大学総合研究博物館研究部所蔵
鳥類は進化の過程で重い歯を捨て去った生物である。そのため食性と嘴の形には密接な関係がある。肉を引き裂く、滑りやすい魚をくわえ上げる、種子を割る、泥の中から小動物をつまみ上げるなど、様々に特殊化した鳥の生態がその鳥に特有な嘴の「形」を生み出した。

052　哺乳類頭骨コレクション
年代未詳／東京大学総合研究博物館資料部医学部門旧蔵／東京大学総合研究博物館研究部所蔵
哺乳類は機能的に分化した異形歯を発達させており、動物の食性は如実に歯に示される。また頭骨は咀嚼のための筋肉と顎関節の配置、感覚器の収納、脳の保護、軽量化など、その種に要求される様々な特徴を備えている。これらの標本は動物が生きるための「形」、進化の果てにある「形」を示してくれる。

053　ムクドリ科 (IMTAc_UT0000026)　1960 年代／骨格標本

054　ニホンザル (IMTAb_UT0000170)
2002年／骨格標本／東京大学総合研究博物館研究部所蔵

055　クリイロリーフモンキー (IMTE_MD0000169)
1904年／骨格標本／ボルネオ／東京大学総合研究博物館資料部医学部門旧蔵／東京大学総合研究博物館研究部所蔵

クロザルもしくはムーアモンキー (IMTE_MD0000170)
年代未詳／骨格標本／東京大学総合研究博物館資料部医学部門旧蔵／東京大学総合研究博物館研究部所蔵

オオカンガルー (IMTAb_UT0000182)
2000年代／骨格標本／東京大学総合研究博物館研究部所蔵

056　オランウータン (IMTE_MD0000171)
1905年／骨格標本／ボルネオ／東京大学総合研究博物館資料部医学部門旧蔵／東京大学総合研究博物館研究部所蔵

057　ペリカン属の一種 (IMTE_MD0000172)
年代未詳／骨格標本／東京大学総合研究博物館資料部医学部門旧蔵／東京大学総合研究博物館研究部所蔵

058　カンガルー属の一種 (IMTE_MD0000034)
1910年／骨格標本／東京大学総合研究博物館資料部医学部門旧蔵／東京大学総合研究博物館研究部所蔵

ヒト (IMTE_KO0000001)
年代未詳／骨格標本レプリカ／東京大学総合研究博物館研究部所蔵

059　ヤギ (雑種)　右上腕骨　2002 年／小笠原村父島 (東京都)／個人蔵
060-062　鳥類骨格標本
063　ハイタカ　2011 年／骨格標本／個人蔵

ミヤコドリ (IMTE_MD0000051)
年代未詳／骨格標本／東京大学総合研究博物館資料部医学部門旧蔵／東京大学総合研究博物館研究部所蔵

064　カラス属 (IMTE_MD0000054)
年代未詳／骨格標本／東京大学総合研究博物館資料部医学部門旧蔵／東京大学総合研

究博物館研究部所蔵

065　ケリ属の一種 (IMTE_MD0000053)
年代未詳／骨格標本／東京大学総合研究博物館資料部医学部門旧蔵／東京大学総合研究博物館研究部所蔵

067　メキシコインコの一種　2010 年／骨格標本／個人蔵
068　セグロカモメ　2012 年／骨格標本／個人蔵

ソリハシセイタカシギ (IMTE_MD0000048)
年代未詳／骨格標本／東京大学総合研究博物館資料部医学部門旧蔵／東京大学総合研究博物館研究部所蔵

069　ニワトリ骨格 (IMTAc_YZ0000003)
年代未詳／骨格標本／静岡県立焼津水産高校旧蔵／東京大学総合研究博物館研究部所蔵

071　シロトキ属の一種 (IMTE_MD0000049)
年代未詳／骨格標本／東京大学総合研究博物館資料部医学部門旧蔵／東京大学総合研究博物館研究部所蔵

ヤマドリ (IMTAc_UT0000013)
年代未詳／骨格標本／東京大学総合研究博物館研究部所蔵

072　ハシブトガラス (IMTAc_UT0000015)
2003年／骨格標本／東京大学総合研究博物館研究部所蔵

073　クビワオオコウモリ (IMTAb_UT0000164)
2004年制作／骨格標本／東京大学総合研究博物館研究部所蔵

074　ホルス像
年代未詳／ブロンズ／東京大学総合研究博物館資料部薬学部門所蔵
ホルスはエジプト神話に登場する天空と太陽を司る神である。本像がハヤブサであるように、猛禽類の姿で表現される。ホルスの右眼は太陽を、左眼は月を象徴し、左眼のかたちがアルファベットの「Rx」に似ていたことから、西洋では処方箋を表すものとして「Rx」マークが用いられてきたという。このことから本像は薬学に縁の深いモチーフとして、東京大学薬学部の建物の装飾に用いられていたものと考えられる。

075　シカ頭骨コレクション
076　大型四肢骨コレクション
077　草食性哺乳類骨格コレクション
078　草食性哺乳類頭骨
079　ウシ (IMTAb_UT0000072)
2003年制作／骨格標本／東京大学総合研究博物館研究部所蔵

ブタ (IMTAb_UT0000014) (IMTAb_UT0000015)
2003年／骨格標本／東京大学総合研究博物館研究部所蔵

080　ヤギ、ヒツジ頭骨
081　ヤギ、ヒツジ、レイヨウ類頭骨
082　ヤギ (IMTAb_UT0000075)
1999年制作／骨格標本／東京大学総合研究博物館研究部所蔵

シバヤギ (IMTAb_UT0000077)
2002年制作／骨格標本／東京大学総合研究博物館研究部所蔵

モウコガゼル (雄) (IMTAb_UT0000076)
2004年制作／骨格標本／東京大学総合研究博物館研究部所蔵

083　カメ類標本

084 スッポン (IMTE_MD0000041)
年代未詳／骨格標本／東京大学総合研究博物館資料部医学部門所蔵

085 ワニ (アリゲーター) (IMTE_MD0000059)
年代未詳／骨格標本／東京大学総合研究博物館資料部医学部門所蔵

086-087 オキゴンドウ (IMTAb_UT0000127)
2008年／骨格標本／横浜八景島シーパラダイスにて飼育／東京大学総合研究博物館研究部制作・所蔵
他のハクジラ類と違い、本種とシャチはマグロや他の鯨類など大型の獲物を襲うこともある。巨大な歯は単に捕らえた獲物を逃がさないだけではなく、大きな獲物から肉を抉り取ることもできる。最大で体長6メートルほどに達し、額部分の窪みはメロンと呼ばれる器官を収納する。この器官はハクジラ類に見られ、高周波音を発振して餌を探知するエコーロケーションに関わると考えられている。

088-089 コキクガシラコウモリ (IMTAb_UT0000038)
2005年制作／東京大学総合研究博物館研究部所蔵

キクガシラコウモリ (雄) (IMTAb_UT0000044)
2003年／福岡県／東京大学総合研究博物館研究部所蔵

オオコウモリの一種 (IMTE_MD0000047)
年代未詳／東京大学総合研究博物館資料部医学部門所蔵

コウモリの一種 (IMTAb_UT0000166)
1968年／剝製標本／東京大学総合研究博物館研究部所蔵

090-091 小型動物骨格標本コレクション　東京大学総合研究博物館研究部所蔵

092 ノウサギ (雌) (IMTAb_UT0000042)
2004年／東京都／骨格標本／東京大学総合研究博物館研究部所蔵

モモンガ (IMTAb_UT0000036)　2006年制作
ヒキガエル (IMTAd_UT0000019)　2006年制作

093 ヨーロッパモグラ (IMTE_MD0000025)
年代未詳／東京大学総合研究博物館資料部医学部門所蔵

094-095 フランス製甲虫標本コレクション
19世紀末から1970年代／2010年購入／乾燥標本／東京大学総合研究博物館研究部所蔵

096 フランス製甲虫標本コレクション
19世紀末から1970年代／2010年購入／乾燥標本／東京大学総合研究博物館研究部所蔵

097 イナヅマツノヤシガイ
年代未詳／産地未詳／乾燥標本／東京大学総合研究博物館研究部所蔵
大型で殻高30センチメートル以上に達し、強く膨らむ。動物体は大きく殻を包み込む。肉食性で、西太平洋熱帯の砂底に生息。卵は体内受精後に卵嚢中に産み付ける。属名の「Melo」はメロン、種小名の「umbilicatus」は臍があることを意味し、メロン形の殻の殻頂部にへこみがある様子を表している。

貝殻標本コレクション
明治10年代／東京大学 (旧) 理学部動物学教室旧蔵／総合研究博物館研究部所蔵
東京帝国大学理学部動物学教室によって収集された貝類標本の代表例。動物学教室では1870 年代以降、様々な動物標本を網羅的に収集したが、そのなかでも貝類は最大の規模をもつ。海外の研究機関との交換によって世界中から標本を集めたほか、関係者を南洋諸島に派遣して熱帯性の貝類も多数収集した。保存箱として使われているのは、ホオ材やサクラ材の薄板で作った経木に、和紙の反故紙を下貼りし、黒漆で塗装したもの。東京帝国大学ではもっとも広汎に使用されていた。

098-099　貝殻標本コレクション
明治10年代／東京大学（旧）理学部動物学教室旧蔵／総合研究博物館研究部所蔵

100-101　マサイキリン（雄）（IMTAb_UT0000150）
2009年／骨格標本／神戸市立王子動物園にて飼育（名前「神平」8歳）／東京大学総合研究博物館研究部制作・所蔵
キリンの首の頸椎は7個で、他の哺乳類と変わらない。キリンの首が長い理由は未だにはっきりしていないが、高い樹上の餌を取るためだけでなく、雄同士の争いに首を用いることが知られている。同時にキリンの脚は極めて長く、現生で最も背の高い動物となっている。体幹部分はごく普通の草食動物の形状をもつ。

104　ミズオオトカゲ（IMTAd_UT0000018）
年代未詳／剝製標本／東京大学総合研究博物館研究部所蔵
東南アジアに広く分布する大型のトカゲ。最大で2メートル30センチに達する。川辺に生息し、遊泳能力がある。肉食性で、昆虫、カニ、魚類、カエル、鳥類を捕食する。

ホラガイ
年代未詳／沖縄か／乾燥標本／東京大学総合研究博物館研究部所蔵
殻は大型で殻長40センチメートル以上に達するが、近年、大型標本の入手は困難である。食性は肉食性でヒトデ類を好んで食べる。特に、サンゴの食害で悪名の高いオニヒトデの天敵として有名である。紀伊半島以南、主に奄美諸島以南の熱帯太平洋に広く分布し、浅海のサンゴ礁域に生息する。

105　東京大学総合研究博物館所蔵・哺乳類剝製標本
東京大学は130年を超える歴史の中で多数の標本を入手して来た。映像技術が未発達であった時代においては、見たこともない動物の姿をそのままに留める剝製標本の価値は非常に高いものであった。これらの標本のなかには現在は珍しいものや、過去の歴史を物語るものも含まれている。アザラシが北海道では珍しいものではなかった時代、タイワンヤマネコが「日本産」であった時代を窺わせる標本である。

老田コレクション・哺乳類剝製標本
岐阜県飛騨市にて老田敬吉・正夫の両氏により運営されていた老田野鳥館（2008年閉館）旧蔵のコレクションである。老田野鳥館は鳥類保全の普及を目的としていたが、郷土自然館としての性格も併せ持っており、哺乳類標本も所蔵していた。ここに収められた老田コレクションは日本人の身近に生息するいわば隣人たち、しかし目にすることは少ない哺乳類たちである。
106　コジャコウネコ（IMTAb_UT0000004）
1976年購入／剝製標本／東京大学総合研究博物館研究部所蔵

107　ノウサギ（IMTAb_OT0000140）
1971年11月19日／（旧）老田野鳥館旧蔵／東京大学総合研究博物館研究部所蔵

ノウサギ（IMTAb_OT 0000143）
年代未詳／（旧）老田野鳥館旧蔵／東京大学総合研究博物館研究部所蔵

ノウサギ（IMTAb_OT 0000144）
年代未詳／（旧）老田野鳥館旧蔵／東京大学総合研究博物館研究部所蔵

108　タイワンヤマネコ（IMTAb_UT0000005）
年代未詳／剝製標本／東京大学総合研究博物館研究部所蔵

アザラシ科の一種（幼体）（IMTAb_UT00002）
年代未詳／剝製標本／東京大学総合研究博物館研究部所蔵

109　イタチ（IMTAb_UT0000007）
年代未詳／剝製標本／東京大学総合研究博物館研究部所蔵

キツネ（IMTAb_OT0000146）
年代未詳／（旧）老田野鳥館旧蔵／東京大学総合研究博物館研究部所蔵

タヌキ (IMTAb_OT0000142)
1968年／(旧) 老田野鳥館旧蔵／東京大学総合研究博物館研究部所蔵

110　アルマジロの一種 (IMTAb_UT0000006)
年代未詳／剝製標本／東京大学総合研究博物館研究部所蔵

111　ツキノワグマ幼獣 (IMTAb_OT 0000149)
年代未詳／(旧) 老田野鳥館旧蔵／東京大学総合研究博物館研究部所蔵

112　ニホンカワネズミ (IMTAb_OT 0000134)
1965年8月6日／(旧) 老田野鳥館旧蔵／東京大学総合研究博物館研究部所蔵

113　ニホンリス (IMTAb_OT 0000135)
年代未詳／(旧) 老田野鳥館旧蔵／東京大学総合研究博物館研究部所蔵

114　ニホンイタチ (雄) (IMTAb_OT 0000121)
1974年採集／岐阜県／(旧) 老田野鳥館旧蔵／東京大学総合研究博物館研究部所蔵

115　リス幼獣 (IMTAb_OT 0000137)
年代未詳／(旧) 老田野鳥館旧蔵／東京大学総合研究博物館研究部所蔵

116　オコジョ (IMTAb_OT 0000117)
年代未詳／(旧) 老田野鳥館旧蔵／東京大学総合研究博物館研究部所蔵

117　ハリネズミ (IMTAb_UT0000008)
年代未詳／東京大学総合研究博物館研究部所蔵

118　モモンガ (IMTAb_OT 0000139)
年代未詳／(旧) 老田野鳥館旧蔵／東京大学総合研究博物館研究部所蔵

119　ウシ解剖模型 (IMTAb_UT0000065)
年代未詳／合資会社山越教育器械標本製作所／東京／紙に彩色／東京大学総合研究博物館研究部所蔵

120-121　ウシ石膏製縮小模型群 (IMTAb_UT0000066, 157-160)
1883-1892 (明治16-25) 年／マックス・ランズベルク製作／ベルリン (ドイツ) ／石膏に彩色／東京帝国大学農学部獣解剖学科および東京大学農学部3号館中央標本室 (動物解剖学教室) 旧蔵／総合研究博物館研究部所蔵
6分の1縮尺のウシ石膏模型。当時のベルリン獣医大学の飼育個体をはじめとし、ドイツの国内外で飼養されていた特定の個体を扱ったスケールモデルであることが記録される。明治期の日本の動物学・畜産学教育はドイツの影響を受け、こうした質の高い教育用標本がドイツから導入された。ジャージー、ショートホーン、シンメンタールの3品種を形態学的に比較できる。

122-124　海産無脊椎動物

125　ゴホウラ
年代未詳／産地未詳／乾燥標本／東京大学総合研究博物館研究部所蔵

126　テングニシ
年代未詳／日本／乾燥標本／東京大学総合研究博物館研究部所蔵

サラサバテイラ
年代未詳／産地未詳／乾燥標本／東京大学総合研究博物館研究部所蔵

シャゴウ　年代未詳／沖縄県西表島／乾燥標本／東京大学総合研究博物館研究部所蔵

127　ミゾコブシボラ
年代未詳／産地未詳／乾燥標本／東京大学総合研究博物館研究部所蔵

クチグロトウカムリ
年代未詳／産地未詳／乾燥標本／東京大学総合研究博物館研究部所蔵

アカニシ　年代未詳／日本／乾燥標本／東京大学総合研究博物館研究部所蔵

テングガイ　年代未詳／産地未詳／乾燥標本／東京大学総合研究博物館研究部所蔵

ヒレジャコガイ　年代未詳／産地未詳／乾燥標本／東京大学総合研究博物館研究部所蔵

128-129　海産無脊椎動物

130　ウニ類標本コレクション
ウニ類は棘皮動物門ウニ綱に属する動物である。体は球形、楕円形または円盤状の殻をもつ。生時は殻の表面が様々な長さの棘に覆われているが、死後は脱落し殻が剥き出しになる。ウニ類は深海から浅海のあらゆる海洋環境に適応しており、日本に約160種、世界に約900種が知られる。岩礁に生息するグループと堆積物中に生息するグループがあり、前者は海藻、後者は堆積物中の有機物を餌とする。

131　カイメンの一種

ヤヨイハルカゼ　年代未詳／産地未詳／乾燥標本／東京大学総合研究博物館研究部所蔵

132　イナヅマツノヤシガイ
年代未詳／産地未詳／乾燥標本／東京大学総合研究博物館研究部所蔵

133　クロチョウガイ
年代未詳／産地未詳／乾燥標本／東京大学総合研究博物館研究部所蔵

134　サンゴ標本コレクション
サンゴは刺胞動物の一種であり、イソギンチャクやクラゲに近縁である。動物体は炭酸カルシウムの骨格を分泌し、サンゴ礁を形成する。大多数のサンゴは、遺伝的に同一な小さな個体（ポリプ）が集まってできる群体である。サンゴ自体はプランクトン食であるが、体内に褐虫藻を共生させており、褐虫藻が光合成により生産する有機物を利用して成長することができる。そのためには十分な光が必要であり、サンゴの生息域は熱帯域から亜熱帯域の浅海に限られている。

135　サンゴ標本コレクション

136　オットセイ（IMTAb_YZ0000002）
年代未詳／骨格標本／静岡県立焼津水産高校旧蔵／東京大学総合研究博物館研究部所蔵

137　地球儀（2800万分の1）
1963年改訂／渡辺雲晴／日本／東京工業大学旧蔵／東京大学総合研究博物館研究部所蔵

138　キノコ模型コレクション　焼津水産高校旧蔵剝製標本

139　キノコ模型コレクション
20世紀前半／国産／紙粘土、木に彩色、ガラス・ケース入／東京大学総合研究博物館資料部医学部門所蔵

140　オオトカゲの一種（IMTAd_YZ0000001）
1933（昭和8）年／剝製標本／静岡県立焼津水産高校旧蔵／東京大学総合研究博物館研究部所蔵

ノコギリザメ（IMTAe_YZ0000001）
年代未詳／剝製標本／静岡県立焼津水産高校旧蔵／東京大学総合研究博物館研究部所蔵

チョウザメ（IMTAe_YZ0000002）
年代未詳／剝製標本／静岡県立焼津水産高校旧蔵／東京大学総合研究博物館研究

部所蔵

141　鉱物・岩石・鉱石コレクション
日本の鉱物学・地質学・鉱山学は、1873（明治6）年に設置された開成学校において、ドイツ人鉱山技師カール・シェンクが講義を行ったことに始まる。開成学校由来のものとして、ドイツのクランツ商会から輸入された多くの鉱物・岩石標本が東京大学に今日も残る。また、シェンクの教えを受けた小藤文次郎や和田維四郎ら、さらに彼らの後継者たちによる積極的な標本収集により、東京大学には膨大な鉱物・岩石・鉱石コレクションが形成された。

142　黄鉄鉱（IMTC_UT0000070）
年代未詳／ナヴァユン、リオハ（スペイン）／個人蔵
鉄の硫化物鉱物。六面体のほかに八面体、五画十二面体、あるいは、これらが組み合わされた外形を示す。鉄と硫黄を完全に分離することが難しく、鉄資源になりにくいため、各地で広範囲に産出されるが、資源的価値は低い。かつては、硫酸を生産するために採掘されたが、硫黄が石油から回収されるようになり、採掘量が減った。多様な外形と生成条件の関係は未詳であり、研究資源としての価値は十分にある。

143　輝安鉱（IMTC_UT0000073）
年代未詳／市ノ川鉱山（愛媛県）／東京大学総合研究博物館資料部鉱物部門所蔵若林標本
アンチモンの硫化鉱物。輝安鉱はアンチモンの主要な原材料であり、古代から化粧用の顔料として用いられた。また、アンチモンは工業材料として広く使用されたが、毒性があると考えられ、その結果、使用されることが少なくなった。鉱物名はアンチモンを表すギリシャ語の「stimmi」あるいは「stib」によって命名された。市ノ川鉱山は輝安鉱の巨大な結晶を産したことで世界的に有名である。

144　石英（水晶：日本式双晶）（IMTC_UT0000075）
年代未詳／乙女鉱山（山梨県）／東京大学総合研究博物館資料部鉱物部門所蔵若林標本
二つの同じ種類の鉱物が、結晶学的に定義できる方位で接合・成長したものを双晶と呼ぶ。石英には多くの種類の双晶が知られるが、本標本のようにV字形（角度は84°34'）で接合した双晶は日本式双晶と呼ばれる。これは山梨県の乙女鉱山から大型の双晶が多数産出したため。水晶は通常六角柱状であるが、日本式双晶の場合は六角板状になる。理由は不明。

145　碧玉（IMTC_UT0000071）
年代未詳／島根県玉湯町／個人蔵
微細なシリカの集合体。不純物によって多様な色彩や模様が現れる。色彩や模様の美麗なものは飾り石として珍重される。

146-147　地球儀（800万分の1）
1937年以前／ベルギー／白地図に彩色／ベルギー政府寄贈／東京大学附属図書館旧蔵／総合研究博物館研究部所蔵
1923（大正12）年9月の関東大震災により、大学の図書館蔵書50万冊が灰燼に帰した。それに対し国内外から援助の手が差し伸べられた。ベルギー政府からは図書と義援金が寄せられ、その善意の記念として、同国に大地球儀の製作を依頼。しかし、大地球儀が贈呈された1937（昭和12）年は日中戦争の最中であり、ベルギーは連合国の一員であったため、震災時に寄せられた友情はすでに過去のものとなっていた。そのためか贈呈された地球儀は「白地図」であった。当時の記録によると「ベルギー国地理学協会側では、特に日本側の好みを考慮し、彩色を施さず白地のままなり」とある。彩色は日本側により、1939年に始められた。

148-149　首長の像（gowe）（70.2001.27.543）
19世紀／ニハ族／ニアス島北西部、リダノ・ラホミ海盆（インドネシア）／石／バルビエ・ミュラー美術館旧蔵／ケ・ブランリ美術館所蔵
ニアス島はインド洋スマトラ島近くに位置し、比較的小規模（120 キロメートル長、40 キロメートル幅）であるが、19 世紀には様々な素晴らしい石造彫刻があることで知られていた。ニアス島の建築、木造彫刻、そして石造彫刻は、様式的に北部・中部・南部の三つの地域に分けられる。起伏のある島の地形がこれら三つの地域を長期にわたり相互に隔ててきたためである。19 世紀末、イタリア人地理学者エリオ・モディリアーニは島の南部地域で、ある首長に関する写真を撮影し、それを持ち帰った。この写真は、頭飾りやそのほかの装飾品がどのように身につけられていたか、またどのように組み合わされていたかをよりよく理解するための貴重な資料となっている。この石像は、胸の上で手を合わせた姿勢で座る首長

を表す。この人物は自分の階級を示すしるしを身につけている。すなわち「ニファタリ」という、もとは金で作られたねじれ状の首飾り、長い耳飾り、ブレスレットである。円錐形の頭飾りの下部にある、額に巻かれた装飾付きのヘッドバンドは、彼の階級に求められる華美で豪奢な装飾がそれに関連した祭宴同様、十分に実現されていることを示す。この人物の仕草は見る人の関心を彼の腰に引きつけるが、この首長が男根により表現される男性性とバランスをとりながら、「人々の母であるがごとく」存在していることを意味している。ニハ族では、勲功祭宴が名誉ある称号の獲得を可能にするため、この祭宴が社会を支配している。この社会的プロセスは村長同士の競争を刺激するとともに、彫像が制作されるための強力な触媒となっていた。ある首長の勲功を祝して作られたこの石像は、先人の存在と精神とを長く記憶に留めさせるものである。

150　キノコ模型コレクション
20世紀前半／国産／紙粘土、木に彩色、ガラス・ケース入／東京大学総合研究博物館資料部医学部門所蔵

151　人体解剖模型（IMTE_MD0001076, 1077, 1078）
石膏に着色／19世紀後半／ドイツ製／東京大学総合研究博物館医学部門所蔵

152　人体解剖模型（IMTE_MD0000184）
年代未詳／石膏に着色／東京大学総合研究博物館資料部医学部門所蔵

153　大澤岳太郎デスマスク（IMTJ_UT0000192）

154　浦上天主堂の獅子頭（IMTC_UT0000100）
1946年5月13日発見／長崎／東京大学総合研究博物館資料部岩石・鉱床部門所蔵
昭和20-21年に行われた原爆被害調査の際に収集された。収集者は調査団員だった渡辺武男東京大学教授。当初は広島護国神社の狛犬として伝えられたが、長崎浦上天主堂の入り口にあったアーチを飾っていた獅子頭であることが判明した。安山岩で作られた獅子頭の頭頂部には原爆の熱線で溶融した安山岩中の黒色鉱物が急冷されてできたガラスが観察される。爆心からの距離は約500メートル。

155　鉱物・岩石・鉱石コレクション

156　巨大ダイヤモンド模型コレクション（IMTC_UT0000101）
1873-1893年／英国製／練ガラス／東京大学総合研究博物館資料部岩石鉱床部門所蔵
ダイヤモンド・レプリカの製造が始められたのは18世紀頃と推定される。それは当時長足の進歩を遂げつつあった自然科学と連動していた。人間にとってダイヤモンドは長く未知の世界のものとしてあり、人々はそれについて様々な伝説を紡いできた。数あるダイヤモンドのなかで、特に名の知られたものを保持することは、根源的には、人類による自然支配を意味している。レプリカとはいえ、19世紀後半の「現状」を留めているという意味で資料価値が高い。

158　ソビエト連邦の代表的な鉱物標本（IMTC_UT0000001）
年代未詳／東京大学総合研究博物館資料部鉱物部門蔵
1964年ケルディシュ博士率いるソ連邦科学アカデミー学術視察団東京大学訪問時の寄贈品。蛍石、ウルトラマリン、トパーズ、オパール、ロードナイト、タルク、玉髄、瑪瑙、辰砂、魚眼石、鶏冠石、ユージアライト、パイライト、サルファー、星葉石、石黄の計16からなるコレクション。

159　欧州産鉱物サンプル標本コレクション（IMTE_MK0000299）
年代未詳／木製箱入り／東京大学医学図書館旧蔵／総合研究博物館研究部所蔵三宅コレクション
三宅秀の父艮斎（1817-1868）が収集し、「一度シーボルトの手に渡ったがその後再び我が国に帰ったもの」との添付文書がある。艮斎はシーボルトの二回目の来日の際に、自ら収集した日本の鉱物の鑑定を依頼した。シーボルトは標本を欧州に持ち帰ったのち、艮斎の要請にもかかわらず返却に応じなかった。秀は父の依頼で欧州訪問の際にシーボルトと面会し、この展示標本を持ち帰った。しかし、これらはボヘミアを中心とした欧州産鉱物の一覧標本で、艮斎の収集標本に代わりシーボルトから受け取ったものと思われる。艮斎の標本は行方不明である。

ロシア産鉱物標本コレクション(IMTC_UT0000002)
年代未詳/東京大学総合研究博物館資料部鉱物部門所蔵
1896(明治29)年に東京大学で鉱物学の教授となった神保小虎(じんぼことら)が、ロシアを訪れた際に持ち帰ったと言われているが、詳細な来歴は不明。

160-161 水晶(IMTC_UT0000023)
年代未詳/福島県石川地方/東京大学(旧)工学部鉱山学科旧蔵/総合研究博物館研究部所蔵

長鼻類右寛骨化石(レプリカ)(IMTD_UT0005012)
東京大学総合研究博物館研究部所蔵

アンモナイト化石(IMTD_UT 0005015)
東京大学総合研究博物館研究部所蔵

162 天球儀(IMTE_MK0000410)
1800年代/ベルリン(ドイツ)/骨木に紙貼り、彩色、方位磁針は金属/東京大学医学図書館旧蔵/総合研究博物館研究部所蔵三宅コレクション
このドイツ製天球儀は日本の三宅家を経て現代へ伝えられたもの。同家は医学分野の歴史において名が知られる医家の一族。

天体望遠鏡(IMTE_MK0000411)
19世紀/P.デルフェル製作/ベルリン(ドイツ)/真鍮/東京大学医学図書館旧蔵/総合研究博物館研究部所蔵三宅コレクション
ヨーロッパの望遠鏡で日本へ伝えられているもののひとつ。本品はドイツ製で、三宅家伝来のもの。同家は帝国大学医科大学初代学長となった三宅秀(1848-1938)をはじめ、特に医学分野の歴史において名が知られる医家の一族である。

163 弥生町出土壺形土器第一号
弥生時代/向ヶ丘弥生町(文京区弥生)/京都科学複製/東京大学総合研究博物館資料部人類先史部門所蔵
いわゆる「最初の弥生土器」とされるもの。現在の東京大学浅野地区周辺の向ヶ岡弥生町(現文京区弥生)において、東京大学予備門生徒有坂鉊蔵らが1884(明治17)年に発見した。当初から「弥生町出土の土器」として知られ、後日、先史考古学の発展とともに「弥生式土器」といった土器形式、さらには「弥生時代」という時代名称の元となった。1975(昭和50)年、国の重要文化財に指定された。本来、割れ口の上部にラッパ形の口縁部が続いていた。

164 石棒コレクション
縄文時代後期から晩期/丸棒状磨製石器/東京大学総合研究博物館資料部人類先史部門所蔵
丸棒状磨製石器の一種。縄文時代前期から晩期に用いられた。一般に剣のかたちを模した祭器と考えられている。東北を中心に北海道から西日本まで広く分布し、形態・大きさ・整形法などによって細分される。展示資料は明治から昭和初期に寄贈・収集され、なかには出土情報が不明なものも含むが、縄文時代後期から晩期(おおよそ4000から3000年前ごろ)の標本と思われる。

シャコガイ貝斧コレクション
年代未詳/ロタ島(マリアナ諸島)/貝殻製道具/東京大学総合研究博物館資料部人類先史部門所蔵
大型のシャコガイの貝殻を加工して斧状の道具としたもの。大正・昭和初期の人類学調査の一環として、マリアナ諸島のロタ島で収集された。貝殻の腹部を利用するのは同諸島に特徴的な型式であり、柄に対し刃部を横方向に据えた手持ちの斧として木彫りに使用されていた。日本では南西諸島の先史時代から主として別型式のものが知られる。

165 石器類参照標本コレクション
19世紀末/ガラス窓付き鋼鉄製ケース入/米国マサチューセッツ州ピーボディ博物館寄贈/(旧)理学部動物学教室旧蔵/東京大学総合研究博物館資料部人類先史部門所蔵
東京大学理学部初代動物学教授の御雇米国人教師E. S. モースは貝類の専門家であったが、1877年に大森貝塚を発掘し、日本で初めて科学的な考古学調査を実践した。モースは

1879年に帰米し、大学が展示場をもつ重要性を説き、神田一ツ橋に「理学部博物場」が設立された。黒枠のガラス標本箱と石器は、モースの計らいにより、「博物場」陳列のために米国ピーボディ博物館から寄贈されたと言われる。同様のものが200箱ほど現存する。

166-167　35ミリ映写機（IMTF0000306）
年代未詳／株式会社富士機械／金属／東京大学（旧）法学部大講堂旧蔵／総合研究博物館研究部所蔵

233　微小貝写真
2012年制作／インクジェット・プリント／1994年標本採集／鹿児島県加計呂麻島沖水深310メートル／乾燥標本原寸4ミリメートル／東京大学総合研究博物館研究部所蔵
写真の微小貝はテンジククダマキ類の一種（軟体動物門腹足綱）。貝類には著しく大型になるものから、小型のものまで、様々なサイズの種が存在する。最大種は二枚貝のオオシャコガイで、大きさは1メートル以上に達する。一方、最小サイズの貝類はミジンワダチシタダミ類で最大で殻径が0.5ミリしかない。1センチメートルに満たない種は多く、特に4ミリメートル以下の小型種は微小貝と呼ばれる。

236-237　鳥類本剝製コレクション
山階鳥類研究所所蔵（東京大学総合研究博物館寄託）
本コレクションの所蔵先山階鳥類研究所は、日本で唯一の鳥類専門の研究機関である。1932（昭和7）年に山階芳麿が自邸に作った資料館を母体とし、質・量ともに第一級の標本コレクションを所有する。総合研究博物館は山階所蔵の鳥類本剝製約350点の寄託を受け、ここに収蔵している。このコレクションには昭和天皇ゆかりの「生物学御研究所」由来の標本が多く含まれており、剝製本体のみならず台座やケースの細工にも、当時の第一級の職人の技を見ることができる。

鳥類本剝製コレクション
1930から1980年代／老田敬吉・正夫収集／（旧）老田野鳥館旧蔵／東京大学総合研究博物館研究部所蔵
岐阜県の老田野鳥館において公開されていた鳥類剝製コレクション。老田野鳥館は老田敬吉氏・正男氏が野鳥の保護思想を啓蒙するために開設した私設博物館であったが、2008年に閉館を余儀なくされた。本館は鳥類を中心とする老田コレクション約300点の寄贈を受け、その一部をここに公開している。

238-239　鳥類本剝製コレクション
山階鳥類研究所所蔵（東京大学総合研究博物館寄託）

240　土佐長尾鶏（赤笹）（IMTAc_WK000079）　高知県／特別天然記念物

241　ハッカン（IMTAc_YS0000634）
1982（昭和57）年／山階鳥類研究所所蔵（東京大学総合研究博物館寄託）

242　コウテイペンギン（IMTAc_YS0000376）
年代未詳／生物学御研究所旧蔵／山階鳥類研究所所蔵（東京大学総合研究博物館寄託）

鳥類本剝製コレクション
1930から1980年代／老田敬吉・正夫収集／（旧）老田野鳥館旧蔵／東京大学総合研究博物館研究部所蔵

243　オジロワシ（IMTAc_YS0000470）
年代未詳／坂本福治制作／生物学御研究所旧蔵／山階鳥類研究所所蔵（東京大学総合研究博物館寄託）

245　天秤ばかり（IMTF0000070）
年代未詳／守谷定吉造／東京／東京大学総合研究博物館研究部所蔵

247　顕微鏡（IMTF0000046）
年代未詳／R.フエス社製／ベルリン／東京大学総合研究博物館研究部所蔵

248　角度計測儀（IMTF0000066）
年代未詳／服部時計店／東京／東京大学総合研究博物館研究部所蔵

249　精密計測機（IMTF0000122）
年代未詳／今井精機製作所／東京大学総合研究博物館研究部所蔵

251　高橋順太郎胸像（IMTJ_UT0000101）
1921（大正10）年／武石弘三郎作／ブロンズ／東京大学（旧）医学部薬理学教室旧蔵／総合研究博物館研究部所蔵
高橋順太郎（1855-1920）は薬理学教室の初代教授。医学部卒業の翌1882年から1885年までドイツに学ぶ。高橋の帰国とともに、薬理学は薬物学の名で独立して教えられることになった。武石弘三郎（1877-1963）は文展で活躍した新潟県出身の彫刻家。東京美術学校彫刻科で長沼守敬に師事し、卒業後、1901年から1909年までベルギー・ブリュッセル王立美術学校に学ぶ。1911年からは文展に出品し、肖像彫刻家として一家をなした。

高橋順太郎胸像（IMTJ_UT0000100）
1921（大正10）年／武石弘三郎作／石膏／東京大学（旧）医学部薬理学教室旧蔵／総合研究博物館研究部所蔵

252　秋篠寺乾漆心木模刻
2008年／菊池敏正制作／檜、漆／個人像
天平時代末期に制作された秋篠寺乾漆心木は、脱活乾漆像の心木のなかでも特殊な構造をもつ作例である。一木造りによる作例が出現し始める時期でもあり、乾漆技法と木彫の関係性を示す構造になっている。角材を差し込む枘穴は非常に多く、角度も様々であり高度な木工技術の基盤があることがわかる。上膊部に一木彫像の背刳りによく似た構造の内刳りがあり、腕の大部分は木心乾漆技法により制作されている。このような特徴から、脱活乾漆技法が衰退する背景には多くの技法が混在していたことを示す作例であり、木心乾漆技法成立の基盤には、木彫の存在が窺える。

254　調速機（IMTF_UT0000005）
年代未詳／シェファー&ブーデンベルク社製／ブーカウ（ドイツ）、マンチェスター（イギリス）／真鍮、鉄、木製台座／東京大学総合研究博物館研究部蔵

フレキシブル・ジョイント（IMTF_UT0000002）
年代未詳／フォン・L. シュレーダー製作／ダルムシュタット（ドイツ）／真鍮、鉄、木製台座／東京大学総合研究博物館研究部蔵

255　材料試験機（IMTF_UT0000147）
年代未詳／エリオット・ブラザーズ社製／ロンドン（英国）／真鍮、鉄、木製台座／東京大学（旧）工学部材料工学科旧蔵／総合研究博物館研究部所蔵
材料の強度を測定するための器具。建築・土木・機械など長らく工学分野の基礎をなしてきたもので、様々なタイプのものが知られる。本品は引張試験のためのもので、計測機構としてばねばかり、負荷機構としてパンタジャッキにあたるものが組み込まれており、試験体へ比較的大きな力を手で加えることができる。

256　東京帝国大学工科大学受賞日英博覧会大賞賞状
1910年／ロンドン（英国）／紙、印刷／東京大学工学部旧蔵／総合研究博物館研究部所蔵
日英博覧会は1902年に結ばれた日英同盟のもと、1910年5月14日から10月29日までロンドンのシェファーズブッシュ19万坪の広大な敷地で開催された国際博覧会。名誉総裁は伏見宮貞愛親王。日露戦争の勝利により、欧米列強と肩を並べたと自負する日本は、英国と対等の近代国家日本を宣伝し、日本製品の対英輸出拡大を図る好機と捉えていた。

257　法医学教室図面

258　幾何学関数実体模型：パラメータ族の放物線が含まれた極小曲面
19世紀末から20世紀初頭／独マルチン・シリング社製／石膏、鉄芯／東京帝国大学理学部数学科旧蔵／東京大学数理科学研究科所蔵
本模型の曲面はパラメータ表示によって「x=a sin 2φ−2aφ+(1/2)a(v^2)sin 2φ+(1/2)bv sin φ, y=−a cos 2φ−(1/2)a(v^2)cos 2φ−(1/2)b cos φ, z=2av sin φ+bφ」と表され

るもので、平均曲率H=0となる極小曲面のひとつである。1パラメータ族の放物線が含まれたもので、カタランの極小曲面（x=a sin 2ϕ−2aϕ+(1/2)a(v^2)cos2ϕ, y=−a cos 2ϕ−(1/2)a(v^2)cos 2ϕ, z=2av sin ϕ）の一般形にあたるもの、あるいはそれらと螺旋面が足し合わされたものでもある。

259　クーズーの角
年代未詳／乾燥標本／個人蔵
クーズーはアフリカ東部から南部にかけて生息する大形のレイヨウ（アンティロープ）で、肩高150センチ、体重300キログラムに達する。角は二回転あまり捩じれながら伸長し、最大で180センチ近くになることもある。おもに疎林に住み、草原には少ない。角の捩れは成長のプロセスからもたらされるものである。角は雄のみにあり、雌を獲得するために発達したと考えられる。

260-263　珍奇物収集キャビネット

264　流体力学的形態（プロペラ模型に基づく）
2011年／菊池敏正制作／檜、漆、真鍮／個人蔵
流体力学実験用のプロペラからイメージされた彫刻。空気抵抗を推力に変換するための機能的な要求が、このような無駄のない「かたち」を生み出している。この表面形状は、たしかに、実験科学の到達点を示すものであるが、それと同時に植物の種子など自然界に存在する「かたち」と通有性をもっている。

265　撮影者未詳　座る日本人女性肖像
年代未詳／原寸縦9.1、横5.9cm／クリスティアン・ポラック氏所蔵より

撮影者未詳　日本人女性肖像
1862年／原寸縦16.1、横12.5cm／クリスティアン・ポラック氏所蔵より

266-267　内田九一撮影　明治天皇肖像
1875年／原寸縦27.0、横18.5cm／東京大学総合研究博物館研究部蔵より

内田九一撮影　昭憲皇后肖像
1875年／原寸縦27.0、横18.5cm／東京大学総合研究博物館研究部所蔵より

撮影者未詳　日本人女性肖像
年代未詳／原寸縦13.9、横9.3cm／クリスティアン・ポラック氏所蔵より

撮影者未詳　二人の日本人女性肖像
年代未詳／原寸縦13.5、横9.0cm／クリスティアン・ポラック氏所蔵より

268-269　ディスデリ　徳川昭武肖像
1867年／原寸縦8.8、横5.5cm／クリスティアン・ポラック氏所蔵より

フレデリック・サットン　徳川慶喜肖像
1867年5月／大坂／原寸縦9.7、横7.7cm／クリスティアン・ポラック氏所蔵より
撮影者未詳　三味線を持つ日本人女性肖像
年代未詳／原寸縦13.9、横9.3cm／クリスティアン・ポラック氏所蔵より

撮影者未詳　日傘を持つ日本人女性肖像
年代未詳／原寸縦8.7、横5.5cm／クリスティアン・ポラック氏所蔵より

270-271　ナダール　河津祐邦肖像
1864年／パリ／原寸縦29.9、横21.3cm／クリスティアン・ポラック氏所蔵より

撮影者未詳　島津忠義肖像
年代未詳／原寸縦14.7、横10.4cm／クリスティアン・ポラック氏所蔵より

撮影者未詳　日傘を差す日本人女性肖像
年代未詳／原寸縦14.4、横9.3cm／クリスティアン・ポラック氏所蔵より

撮影者未詳　日本人女性肖像
年代未詳／原寸縦13.9、横9.3cm／クリスティアン・ポラック氏所蔵より

272　赤瀬川原平　大日本零円札発行所『零円札』
1969年／紙の両面に活版印刷／個人蔵
お札は印刷物である。何万枚、何十万枚という単位で、一部の狂いもなくマス複製されたものが「真札」である。こうした貨幣経済システムの上に成り立つ、現代の日常的メカニズムの虚をついた赤瀬川原平の「模型千円札」は、1965年「通貨及証券模造取締法」違反にあたるとして起訴され、1970年に最高裁で有罪判決を下された。本品は赤瀬川が考案した「本物」の「零円札」であり、まさに「概念芸術」と呼ぶに相応しい。現在の美術マーケットでは「零円札」が高値で取り引きされており、市場はそれを歴史的に価値ある美術品として認定している。

ワイマール共和国インフレ紙幣コレクション
1914年から1923年／ドイツ／紙に印刷／東京大学総合研究博物館研究部所蔵
第一次世界大戦後のドイツでは、10年も経ぬ間に極端なインフレが進行した。ここには1914年の1マルク紙幣から1923年の5億マルクまでが並ぶ。

273　貨幣コレクション

274　幕末医家三宅一族コレクション
19世紀から20世紀初頭／東京大学医学図書館旧蔵／総合研究博物館研究部所蔵
三宅秀 (1848-1938) を中心とする近代医家三宅一族の旧蔵コレクション。三宅秀は日本初の医学博士の一人。東京大学医学部教授、医学部長、名誉教授、貴族院議員などを歴任した。1863 (文久3) 年から翌年にかけて、徳川幕府の遣欧使節団に最年少の随員として加わった。その際に購入された医療器具や学術標本が多く残されている。その後も秀は海外出張の度に、外科道具や科学機器など、様々な外国製品を購入していた。

275　六連発拳銃 (IMTE0000415)
19世紀後半／ベルギー／鉄、木製握把、木製箱
三宅秀の父艮斎 (1817-1868) 旧蔵の拳銃。銃身部に「九年一四五七東京府」と刻字があり、明治9 (1876) 年に東京府によって登録番号が付された。

276　幕末医家三宅一族コレクション
19世紀から20世紀初頭／東京大学医学図書館旧蔵／総合研究博物館研究部所蔵

277　分銅 (IMTE0000355)　　金属、木製箱

278　胃鏡 (IMTE0000336)　　金属、革製箱

279　外科道具セット (IMTE0000307)
ライター製／ウィーン (オーストリア) ／金属、鼈甲、革製ケース

281　遠近法実体模型 (IMTG0000108)
1882年／ハーレ社製／ドイツ／石膏に鉄線／東京帝国大学理学部数学科旧蔵／東京大学数理科学研究科所蔵

282　ヘチマ　年代未詳／台湾／個人蔵

283　ダチョウ卵殻　年代未詳／個人蔵
ダチョウの卵は現生の鳥のなかで最大であるが、体重に対する割合でいえばもっとも小さい。体に見合わぬ小さな卵をたくさん産むように進化した鳥。

284　数理科学教育教材
19世紀後半／木／東京帝国大学理学部数学科旧蔵／東京大学数理科学研究科所蔵

285　宝石珊瑚
年代未詳／産地未詳／乾燥標本／東京大学総合研究博物館研究部所蔵
サンゴとは刺胞動物のうち硬い骨格を分泌する種の総称であり、そのうち宝石として利用さ

れる種は宝石珊瑚と呼ばれる。動物体は多数のポリプの群体からなり、炭酸カルシウムの骨格を覆う。骨格は枝状に分岐し、色彩は赤色で美しい。深海の岩礁に生息する。

286　自然金モデル（IMTC_UT0000093）
19世紀／クランツ商会／ドイツ／石膏／東京大学総合研究博物館資料部岩石・鉱床部門所蔵クランツ標本
1842年ロシア、ミアスクで採掘された当時世界最大の自然金（砂金、約40キログラム）のレプリカ。ドイツの著名な鉱物標本商であるクランツ商会において製造された。レプリカではあるが、掘り出されたときの「かたち」を留めるものとして貴重である。

287　鉄電気石（IMTC_UT0000096）
年代未詳／ゴルコンダ鉱山（ブラジル）／東京大学総合研究博物館資料部鉱物部門所蔵
自然界には驚くべきフォルムが溢れている。なかでも、鉱物界は多様性に富む。鉄電気石は「電気石(トルマリン)」グループに属し、鉄分のために黒色となる。主に花崗岩(ペグマタイト)に産する。熱を加えるか、圧力をかけると、分極が現れることから、「電気石」と呼ばれるようになった。トルマリン・グループの中でも、リチウムを含むトルマリンは、美しい多様な色を示すため、宝石として珍重される。

288　球状方解石（IMTC_UT0000095）
年代未詳／月布鉱山（山形県）／東京大学総合研究博物館資料部鉱物部門所蔵
ベントナイト中に見られる球状の方解石。しばしば温泉析出物として見られる。

289　黒漆稜花合子
中国 元時代、14世紀／木胎、漆／田中儀一旧蔵品
十弁の稜花形からなる合子。蓋上面にはタイマイの板が嵌められていたと推測される。合口や花弁の輪郭部分には鉛で縁取りを施している。北宋時代に多く作られた端正な形と漆の色の美しさを求める無文漆器から、堂々とした力強い表現へ移行する過渡期の作品である。

290　宮中下賜品紋入ボンボニエール
銀製／田中儀一旧蔵品
フランス語で「小型のキャンディー缶やつぼ」を意味するボンボニエール（Bonbonnière）に由来する金工品。結婚式や宗教的なお祝いの儀式等の特別な機会に招待客に振る舞うプレゼントとして知られる。日本でも皇室がそのようなヨーロッパの伝統を引き継ぎ、皇室御慶事の引き出物として菊花紋を入れ、使うようになったと言われる。銀は明治20年代の国際社会の中で紙幣と同価値の貴金属としての値打ちをもち、欧米では王室外交に銀器が多く用いられていた。ボンボニエールが引出物として出された儀式は、御誕生、御着袴、成年式、立太子、御結婚、大礼（御即位）、御結婚25年（銀婚式）を主とし、御在位の記念、海外御訪問の記念、長寿の御祝等である。その他、外国の皇族や大使との午餐にも出された。展示品7点のうち、明治天皇女子御慶事記念品3点には、何の折に下賜されたものかについて田中林太郎が記した書付がある。また、大正大礼大饗記念品、昭和天皇の立太子および成年式記念品の3点には、それぞれ宮内大臣波多野敬直から差し出された田中不二宛の饗宴招待状が残っている。

291　六葉花弁形香合型鴛鴦文
1916（大正5）年11月27日／迪宮裕仁親王（昭和天皇）立太子記念／刻印「純銀　小林製」

292　無題（東宮御所御写真）
1909（明治42）年／小川一眞撮影／2帙154枚、コロタイプ印刷、黒革製帙、木製外箱／田中儀一旧蔵品
東宮御所（現迎賓館赤坂離宮、国宝）竣工時の写真。皇太子嘉仁親王（大正天皇）の住居として建設されたネオ・バロック様式の欧風宮殿建築。鉄骨煉瓦造、外壁石造。設計は片山東熊。当代一流の建築家、美術家、技術者等が結集し、10年もの歳月をかけて完成した。写真は建物外観、各室内装、浴室、廊下、地下設備など建物全体を網羅する。田中林太郎は当時最先端の暖房設備関連業務への尽力により本写真を下賜された。宮内庁書陵部には本写真とほぼ同寸のゼラチン・シルヴァー・プリントの写真帖が所蔵されており、白革製帙の上製版と黒革製の並製版があり、皇室に献上されたと推察される。本コロタイプ版は写真師小川一眞自身が製作し、関係者に頒布されたものであろう。小川は翌1910年に写真師として初めて帝室技芸員を拝命した。

東洋文庫読者倶楽部

発会のご挨拶

東洋文庫は1963年10月に創刊いたしました。爾来50年にわたり、日本・韓国・中国・インド・イスラム圏を含む広大な地域でアジアの先人たちが築いた価値ある古典を発掘し続け、これまでに約800巻を刊行してまいりました。2013年秋、創刊50周年を迎えるにあたり、ファンクラブ「東洋文庫読者倶楽部」を発足させていただきます。東洋文庫を日頃からご愛顧いただいている読者と著者、出版社をむすぶ交流の場として多くの皆さまの入会をお待ちしております。

入会特典　［入会費無料］

■『東洋文庫マイブック』プレゼント　（入会先着300名様）

東洋文庫と同じ装丁・ケース入りの『東洋文庫マイブック』をプレゼントします。ページには縦の罫線のみ入れています。日記や読書記録など、あなただけの「東洋文庫」としてお使いください。

■ 会報誌『東洋文庫通信』発行

777円 840円 777円 1260円 1785円 3360円 3045円 6300円 2940円 3360円 2940円 1575円 1890円 2940円 2940円 10500円

白川静を読むときの辞典

編＝立命館大学白川静記念東洋文字文化研究所

白川静の文章に頻出する、呪禁・載書・卜辞・説文などの語。これらの白川学をより深く知るために必要な用語や作品名を解説する。あわせて白川静主要著作も紹介。項目数約600。

1890円

絵入簡訳 源氏物語 一（全3巻）

小林千草、千草子（センソウコ）

江戸時代のベストセラー、山本春正の『絵入源氏物語』より貴重な挿絵をちりばめ、適宜簡訳により現代語訳した新しい源氏物語の世界。主要な和歌は全て収め、絵と歌で楽しめる一冊。

2730円

金井美恵子エッセイ・コレクション［1964-2013］3 小説を読む、ことばを書く

金井美恵子

吉田健一、石川淳、大岡昇平、深沢七郎、吉岡実、澁澤龍彦、山田風太郎、石井桃子、中上健次など、多彩な作家たちのテクストを読む。解説エッセイ＝朝吹真理子、推薦文＝笙野頼子

予価2310円

包丁侍 舟木伝内 加賀百万石のお抱え料理人

陶智子、綿抜豊昭

舌の肥えた藩主をうならせた加賀藩お抱え料理人、舟木伝内。江戸時代に実在した料理人の仕事と実像とは。12月14日全国公開の映画「武士の献立」関連作品。

1680円

千本組始末記 アナキストやくざ笹井末三郎の映画渡世

柏木隆法

ギロチン社、大杉栄らとともにアナキズムと深く関わりながら〝かたぎやくざ〟として生き、日本映画史の揺籃期から全盛期を駆け抜けた侠客・笹井末三郎の生涯を描くノンフィクション。

3990円

地図で学ぶ日本の歴史人物

監修＝池上彰

卑弥呼から山中伸弥まで総勢300人！47都道府県にまつわる重要人物を通して、日本の歴史と地図を学ぶ。自分の住む県から隣県へ興味が広がる、ユニークな「地図＋図鑑」の連係学習。

1995円

SUNAO SUNAO 3

100%ORANGE

人気イラストレーター、100%ORANGEの本格コミック・シリーズ、第3弾。可愛くてポップな主人公・スナオ君が今度もパンクにシュールに大活躍!? 「ウェブ平凡」で話題連載の待望の書籍化！

予価1050円

愛蔵版 銀曜日のおとぎばなし1（全3巻）

萩岩睦美

イギリスを舞台に小人のポーが活躍するファンタジーコミックの金字塔。「りぼん」連載時の本文カラーをすべて復活し、著者による30年ぶりの描き下ろしマンガも掲載。 デザイン＝名久井直子

1260円

京都の喫茶店 昨日・今日・明日

木村衣有子

京都カフェブームの火付け役となった『京都カフェ案内』から12年。改めて選びぬかれた15軒から見えてくる、いままでとこれからの京都の喫茶店。特別対談＝鷲田清一

1470円

決定版 世界のパン図鑑224

監修＝大和田聡子

世界のパン約230種をカラーで紹介。既刊類書にはない数と種類を網羅し、地域・国ごとに収録。それぞれのパンの特徴、材料の配合割合、飲み物などとのマリアージュも記載！

1785円

生きられた家をつむぐ

くうねるところにすむところ：家を伝える本シリーズ32

奥山明日香

建築の解釈に多くの影響を与えつづけている思想家多木浩二の著作『生きられた家』に触発され、シェアハウスの日常性をみずみずしい感性でつむぎだした、「建築女子」による珠玉の家の本。

予価1995円

縄文の力

別冊太陽 日本のこころ212

監修＝小林達雄

奇想的で過剰とも見えるほど独特の形態をした土器、森羅万象が息づく精霊を宿した不思議な土偶たち、聖なる場に造られた遺跡の数々。現代人の心を揺さぶる縄文パワーの謎に迫る！

2730円

郷土菓子 ふるさとの味を旅する

別冊太陽の地図帖22

編＝太陽の地図帖編集部

日本各地の風土と歴史が育んできた伝統の味。全国のお菓子紹介、日菓と行く京都和菓子散歩、シュガーロード探訪など郷土菓子最前線が満載。文＝酒井順子、中山圭子、若菜晃子ほか。

1260円

坑夫・山本作兵衛の生きた時代 戦前・戦時の炭坑アート

コロナ・ブックス185

編＝作兵衛事務所＋ポレポレタイムス社

ユネスコ世界記憶遺産に認定された山本作兵衛の炭鉱絵画。その作品群を中心に、作兵衛以外の表現者による絵画、写真、ポスターなども併せて紹介。美術で巡る日本の炭鉱。

1785円

作家の家2

コロナ・ブックス186

川本三郎、小山明子、斎藤由香ほか

作家が長年暮らした住まいを紹介。玄関、書斎、リビング、庭など、作家ならではのスタイルを探る。永井荷風、川端康成、北杜夫、塔本シスコ、武満徹、大島渚、安井かずみ、藤田嗣治ほか。

予価1680円

こころ Vol.15

半藤一利、村田喜代子ほか

徹底討論「世界文明と日本文化」＝高階秀爾＋芳賀徹＋山崎正和＋三浦雅士／半藤一利「日露戦争史」、いよいよ日本海海戦！／村田喜代子、落合恵子、池澤夏樹、武田花ら好評連載。

840円

※定価の表示は本体価格に消費税（5%）が加算されています。
本のご注文はお近くの書店へ。小社直接の場合は、平凡社サービスセンター 0120-456987またはホームページのオンラインショップをご利用ください。

293 独マルチン・シリング社製幾何関数実体模型コレクション
19世紀末から20世紀初頭／石膏（東京帝国大学理学部数学科旧蔵、東京大学数理科学研究科所蔵模型より複製）／東京大学総合研究博物館研究部所蔵
模型はそれぞれ、代数曲面とその特異点の構造を扱った代数幾何学に関するもの、定曲率曲面、極小曲面などを扱った微分幾何学に関するもの、楕円関数など複素関数論に関するものに分けられる。これらは、いわゆる概念模型ではなく、数値計算に基づく精密なものである。複雑な曲面の模型化は容易ではないが、それらが一定の精度のもとに実現されている点でも貴重である。東京大学数理科学研究科に収蔵されているオリジナル模型は、中川銓吉教授により20世紀初頭に輸入され、当時は、理学部数学教室の授業にも用いられていた。1932年にマルチン・シリング社は需要の減少から模型制作を中止している。

294-296 数理模型

297 ジーフェルト曲面―正の定曲率をもつ曲面（IMTG_UM0000114）
石膏（レプリカ）／2011年／菊池敏正制作／東京大学総合研究博物館研究部所蔵

298 正の定曲率をもつ一般化されたヘリコイド曲面（IMTG_UM0000070）
石膏（レプリカ）／2011年／菊池敏正制作／東京大学総合研究博物館研究部所蔵

299 特異点をもつ曲面（IMTG_UM0000113）
石膏（レプリカ）／2011年／菊池敏正制作／東京大学総合研究博物館研究部所蔵

304 徳川武定制作船型模型および流体実験用具コレクション（IMTF0000132）
昭和初期／木／東京大学（旧）工学部船舶工学科旧蔵／総合研究博物館研究部所蔵
主として船体の形状と推進抵抗の相関性について実験的に検証するために作られた模型。徳川武定（1888-1957）ゆかりのものとして東京大学船舶工学科へ伝えられた。海軍技術中将、海軍技術研究所長、東京帝国大学教授を歴任した徳川武定は、水戸藩最後の藩主となった徳川昭武の次男にあたる。松戸徳川家の始祖となった。

305 徳川武定制作船型模型および流体実験用具コレクション（IMTF0000132）
昭和初期／木／東京大学（旧）工学部船舶工学科旧蔵／総合研究博物館研究部所蔵

306 江上波夫コレクション
東京大学名誉教授江上波夫（1906-2002）は、日本国家の起源に関わる騎馬民族征服王朝説の提唱で知られる。あるいはユーラシア大陸各地で実に70年間ほども続けられた広範な歴史・考古系フィールドワーク主催者としても名高い。ユーラシア人文科学の泰斗であり、1991年には文化勲章を受章した。スケールの大きな研究者であったと同時に、江上は無類の考古、歴史、美術作品のコレクターでもあった。半世紀以上にもわたる野外調査の機会をとらえて収集した作品は、ユーラシア各地の考古、歴史、美術を包括的に伝える一大コレクションをなしている。展示品は教授の没後、御遺族から寄贈された品々の一部である。

307 古代ペルシャの首飾り（IMTH_EG_000045）
東京大学総合研究博物館資料部考古部門所蔵江上波夫コレクション
ヒトが装飾品を身につける行為はホモサピエンスの登場以降、10万年前には始まっていた。当初は貝殻や貴石、骨角など自然物が用いられたが、新石器時代以降になると一気に人工的な装飾品作成技術が進展した。古代の西アジアで特徴的な発達を遂げたのは石英で作ったファイアンスである。展示品のなかの薄く緑がかった青色ビーズがそれである（濃い緑はガラス）。こうした装飾品の多くは墓の副葬品として見つかる。被葬者の生前の地位や出自を反映していることが多い。

環状青銅製品
紀元前8-6世紀／イラン／東京大学総合研究博物館資料部考古部門所蔵
イラン鉄器時代の青銅製鋳造品の一種。確たる用途は不明だが、インゴットとされることが多い。本例は全周が閉じることはなく、環部中央付近に厚みをもつ点に特徴がある。端部にはしばしば綾杉文やジグザグ文が施される。1960年代初頭に東京大学の調査団がイラン北部で入手した。

308 古代ペルシャの首飾り（IMTH_EG_000046）
東京大学総合研究博物館資料部考古部門所蔵江上波夫コレクション

309　古代オリエントのガラス（IMTH_EG_0000034、0000036、0000037）
ローマ時代以降／東京大学総合研究博物館資料部考古部門所蔵
ガラス製作のルーツは紀元前2500年頃の古代メソポタミアにさかのぼる。当初の作品は小さなビーズや印章が中心であったが、前1500年頃には容器が作られるようになった。さらにローマ時代に吹きガラス技法が開発されるとガラス工芸は一挙に豊かなものになる。ローマ時代に続くサーサーン朝（3-7世紀）の作品が比較的厚手なのは、長距離交易用という説がある。一番右は、シルクロードを経て正倉院にまで運ばれたサーサーン朝カットガラスと同工の作品である。

310　古代オリエントのガラス（IMTH_UT_000016）
ローマ時代以降／東京大学総合研究博物館資料部考古部門所蔵

311　コブウシ型土器（IMTH_EG_000033）
紀元前1000年前後／イラン北西部／東京大学総合研究博物館資料部考古部門所蔵江上波夫コレクション
イラン北西部、カスピ海沿岸で収集された。肩の上が盛り上がったアジアの畜牛、コブウシを模している。口が突き出ているのは、それ以前からイラン北西部で流行していた嘴型注口土器の伝統を引くもの。耳の破損部を見るとイヤリングが付けられていたとみられる。コブウシ形土器が見つかるのは、もっぱら墓である。葬送儀礼に用いられたことが推測できるが、何を入れた容器であったのかなど詳細は解明されていない。

312　イランの古代土器（IMTH_UT_000022）
先史時代末期（紀元前3千年紀末）から歴史時代／イラン北部／東京大学総合研究博物館資料部考古部門所蔵
東京大学の考古学調査隊が収集した。動物をモチーフにした装飾が目立つ。長い注口をもつ土器は鉄器時代のもので、おそらくトリの嘴を模したもの。

313　古代イランの青銅剣（IMTH_UT_000010、000014）
紀元前1000年前後／イラン北部／東京大学総合研究博物館資料部考古部門所蔵
イラン北部産の青銅製剣。中近東では前2000年紀末には鉄器時代が始まるが、同じ頃、青銅器の製作も続けられていた。ほとんどが墓の副葬品として見つかる。武器として十分使える剣ではあるが、むしろ権威のシンボルとして意味をもっていた可能性が高い。1950年代に東京大学の考古学調査隊が収集した。

314-315　ニワトリ本剝製コレクション
1900年代後半／日本農産工業株式会社収集／和鶏館旧蔵／東京大学総合研究博物館研究部所蔵
日本農産工業株式会社が所蔵し、同社内の和鶏館で公開していたニワトリ・コレクション。日本においても各地域で特色あるニワトリの品種が数多く作出されており、その目的は闘鶏用、食用、鳴き声を楽しむため等、多様である。矮鶏（チャボ）は日本独特の愛玩用品種として知られる。これらの品種には天然記念物に指定されているものも多い。

316　猩々蓑曳（雄）（IMTAc_WK000033）　愛知県、静岡県／天然記念物
317　髯鶏（IMTAc_WK000021）　新潟県佐渡島
318　蓑曳矮鶏（尾曳）（IMTAc_WK000041）　高知県／天然記念物
319　銀笹矮鶏（IMTAc_WK000074）

320　声良（雄）（IMTAc_WK000030）　秋田県／天然記念物
321　真黒矮鶏（IMTAc_WK000053）
322　白牡丹矮鶏（IMTAc_WK000046）
323　白桂矮鶏（IMTAc_WK000076）

324　褐色鶉尾（IMTAc_WK000018）　高知県／天然記念物
銀波矮鶏（IMTAc_WK000050）

325　白軍鶏（IMTAc_WK000049）　天然記念物

326　巨大昆虫標本江田茂コレクション
主に20世紀後半／乾燥標本／東京大学総合研究博物館研究部所蔵

江田茂 (1930-2008) は世界の昆虫を集めた大収集家である。この巨大で奇妙な昆虫は総合研究博物館に寄贈された膨大なコレクションの一部である。現在の地球環境では昆虫の大きさに制約がある。それは構造や重量、天敵、そして呼吸能力に関わる。ここに展示した昆虫のうち、ヨナグニサンは日本最大の鱗翅類であり、コノハムシは進化が生んだ驚異そのものである。

327 ナナフシの一種 (IMTAf_ED0001294)

ヨナグニサン (雄) (IMTAf_ED0001274)
ヨナグニサン (雌) (IMTAf_ED0001275)
ヨナグニサン (繭) (IMTAf_ED0001276)

328 アカネクマゼミ (IMTAf_ED0001271)

アユタヤゼミ (IMTAf_ED0001272)
モエギクマゼミ (IMTAf_ED0001273)

329 セレベスコノハムシ (IMTAf_ED0001287)

オオコノハムシ (IMTAf_ED0001289)
セレベスコノハムシ (IMTAf_ED0001288)

330-331 レオナルド・ダ・ヴィンチによる飛行機械 (縮尺20分の1)
1996年／ジョヴァンニ・サッキ制作／木、キャンバス布、糸、革に彩色／個人蔵
ミラノのレオナルド・ダ・ヴィンチ国立科学技術博物館は、レオナルドの構想した各種の機構・機械の模型復元の展示でよく知られている。その多くを制作したのが木工職人ジョヴァンニ・サッキである。翼を拡げた機械の基本的な形状は、その後、英国人ジョージ・ケイリー卿の人力飛行機、リリエンタール兄弟のグライダー、ロシア構成主義者ウラジーミル・タトリンの人力飛行機「レタトリン」にまで受け継がれていく。「飛翔」にかける人類の夢、その第一歩を踏み出したのはレオナルドであった。

レオナルド・ダ・ヴィンチ　トリノ王立図書館所蔵『鳥の飛翔に関する手稿』
1505年3-4月頃／1979年／岩波書店／日本版ファクシミリ／東京大学総合研究博物館研究部所蔵
本手稿には、第18葉表から第4葉裏にかけて「鳥の飛翔に関する論考」が収められる。谷一郎・小野健一・斉藤泰弘の三氏の解説によると、レオナルドの言う「鳥」は、有翼動物と、自ら構想した飛行機械の両方を指しているという。この後者すなわち「巨大な鳥」と呼ばれるものには、バネ動力式、蝙蝠型翼式、人間腹這い式があった。レオナルドは1503年から3年間にわたるフィレンツェ時代に、郊外の丘陵でミミズクをはじめとする猛禽類の飛翔をつぶさに観察し、その観察結果を人間飛行計画に活かそうと考えた。

332-333 インドクジャク尾羽 (白化個体)、ヤマドリ尾羽、インドクジャク三列風切羽、キジ尾羽、猛禽類初列風切羽
2012年制作／インクジェット・プリント／ (旧) 老田野鳥館旧蔵羽毛標本／東京大学総合研究博物館研究部所蔵

オオフウチョウ
2012年制作／インクジェット・プリント／ (旧) 老田野鳥館旧蔵剥製標本／東京大学総合研究博物館研究部所蔵

334 鳥類本剥製コレクション
1930から1980年代／老田敬吉・正夫収集／ (旧) 老田野鳥館旧蔵／東京大学総合研究博物館研究部所蔵

335 フクロウ (IMTAc_OT0001280)
(旧) 老田野鳥館旧蔵／東京大学総合研究博物館研究部所蔵 (同336-345)

336 サンコウチョウ (雄雌) (IMTAc_OT0001053, 0001054)
337 アカショウビン (IMTAc_OT0001153)

338 シマハラヤイロチョウ (IMTAc_OT0001177)
339 カンムリワシ (IMTAc_OT0001308)
340 カワアイサ (IMTAc_OT0001071)
341 アカゲラ (IMTAc_OT0001116)

342 ヤマドリ (雄) (IMTAc_OT0001044) 1981 (昭和56年)
343 コサギ (冬羽) (IMTAc_OT0001073) 1980 (昭和55年)
344 サシバ (雌) (IMTAc_OT0001036)
345 オオコノハズク (IMTAc_OT0001127)

346 トキ (IMTAc_YZ0000001)
年代未詳/静岡県立焼津水産高校旧蔵/東京大学総合研究博物館研究部所蔵
1965 (昭和40) 年以前より静岡県の高校が所有していたもの。当時はまだ生存していた日本産トキと推測される。

347 フラミンゴの一種 (IMTAc_OT0001201)
(旧) 老田野鳥館旧蔵/東京大学総合研究博物館研究部所蔵 (同348-356)

348 アオサギ (IMTAc_OT0001070)
349 ベニヘラサギ (IMTAc_OT0001187)
350 キジ (オス) (IMTAc_OT0001040) 1980 (昭和55年)
351 オシドリ (雄) (IMTAc_OT0001128)
352 ヨーロッパオオライチョウ (IMTAc_OT0001200)

353 オオヅル (IMTAc_UT0000018)
年代未詳/東京大学総合研究博物館研究部所蔵

354 ドンソンゴニ (狩人のハープ) マリ共和国ブグニ周辺/マンデ系バンバラ族
355 バラ (バラフォン) マリ共和国キタ周辺/マンデ系マリンケ族
356 ケセケセ (カシシ) マリ共和国/マンデ系部族
357 タマ (トーキング・ドラム) マリ共和国ニョロ周辺/マンデ系部族
358 ンゴニ (アフリカン・リュート) マリ共和国フータ・トロ周辺/マンデ系部族
359 カリニャン (ギロ) マリ共和国/マンデ系部族

360 製図器セット (IMTF0000129)
1876 (明治9) 年2月/大日本工部省工学寮工作所/黄銅、象牙、エボナイト、木、杉材製ケース入り/東京大学大学院工学系研究科産業機械工学専攻設計工学研究室旧蔵/総合研究博物館研究部所蔵
明治最初期の工部省工学寮時代の銘の存在を確認できる極めて貴重な工学遺産。工部省工学寮は、欧米から工業技術を導入するため、工部省の一等寮として設立が認められた工業教育機関である。ヘンリー・ダイヤーら英国人教師を迎え、1873 (明治6) 年8月に、予科2年、専門科2年、実地科2年の計6年を修業年とした全寮制のエンジニア教育が開始された。「明治九年二月」の年記は第一期生が予科を終え、専門科に進学した頃にあたる。製図学は、機械工学や造船学の予科と専門の基礎学であった。

361 メジャー (IMTF0000131)
年代未詳/金属/東京大学大学院理学系研究科地理学教室旧蔵/総合研究博物館研究部所蔵

362 工部大学校機構模型コレクション
1870年代/真鍮、鉄、木製台座/東京大学 (旧) 工学部機械工学科旧蔵/総合研究博物館研究部所蔵
様々な要素を介した動力伝達の仕組みがわかりやすく表現された模型。明治初頭から機械工学の教育に使われてきた。ねじ機構、てこ・クランクやスライダなどからなるリンク機構、歯車機構などいずれも基本的な内容のものである。一部に1874 (明治7) 年と1875 (明治8) 年の金属プレートの附されたものがあり、国内に残されている西欧近代の産業機械工学の遺産として、最古級に属する。どれも英国製であることから、1873 (明治6) 年に都検として工学寮へ赴任してきたヘンリー・ダイヤーが工学関連の書籍資料類とともに母国から取り寄せたものであろう。

363 三角カム (IMTF0000004)
364 差動ねじ (IMTF0000011)
365 差動歯車機構 (IMTF0000013)

366 早戻り機構 (IMTF0000032)
フォン・シュレーダー社製／ダルムシュタット (ドイツ) ／金属

367 機構モデル (IMTF0000023)　1875 (明治8) 年
368 四リンク機構 (IMTF0000015)
369 ウォーム歯車機構 (IMTF0000018)

370-371 ラチェット (IMTF0000010)
機構モデル (IMTF0000024)

372 金属切削早戻り機構 (IMTF0000014)
373 機構モデル (IMTF0000031)

374 回転斜板機構 (IMTF0000012)
グスタフ・フォイクト機械製／ベルリン (ドイツ)

375 内かみあい平歯車機構 (IMTF0000019)
376 外かみあい平歯車機構 (IMTF0000007)
377 歯車を用いた往復運動機構 (IMTF0000009)

378 ハートカム (IMTF0000001)　1874 (明治7) 年
原節 (カム) の回転運動に対して従節が規則的に従動する、カム機構と呼ばれる基本的な動力伝達機構のひとつ。本模型は明治初頭工部省工学寮において教育用の教材として使われていた。カムの回転中心から端部までの距離が角度に比例していることから、原節 (カム) の角速度が一定のとき、従節の速度が一定となる。

379 温度計
380 タイプライター (IMTF0000116)　ロイヤル社製

381 モダマ豆果 (IMTB_UT0000596)
2005年／ミャンマー／乾燥標本／東京大学理学部植物標本室旧蔵／総合研究博物館研究部所蔵
モダマは、熱帯・亜熱帯に生え、幹の直径が30センチメートルにもなる巨大なつる性の豆の一種で、長さ1メートルにもなる木質の豆のさやをつける。アジア、オーストラリア、アフリカに分布し、日本では九州南部の屋久島以南の海岸近くに生える。種子は水に浮き、海流に乗って広く散布される。和名は「藻玉」の意で、海岸に漂着した種子を海藻の種子と誤解したためと言われる。種子はあずき色で丸く、径5センチメートルほど。細工ものにも使われる。

382-383 ヒマラヤの植物
ヒマラヤの高山帯は、長い冬と短い夏、花期に毎日降り続く雨、晴れたときには強い紫外線と、植物にとっては過酷な環境である。そのような厳しい環境下で生長し、花を咲かせ、子孫を残す植物には、洗練された美しさがある。1960年代以降、ヒマラヤ地域の植物の多様性を解明すべく、東京大学が中心となって多くの調査隊をヒマラヤ地域に派遣してきた。その結果、この50年の間に数多くの知見が得られ、研究の基礎となる標本も多数蓄積された。

ジョン・ジェームス・ラフォレスト・オーデュボン　『アメリカ産鳥類図譜』(複製)
1830-1839年／私家版／ロンドン (英国) ／紙に多彩色石版／東京大学総合研究博物館研究部所蔵
英海軍艦長の庶子として西インド諸島に生まれたオーデュボン (1785-1851) は、自らの足を使って北米大陸各地を旅行し、鳥類を捕獲し、剥製を作り、19世紀鳥類図譜の金字塔とされる本図譜を残した。多彩色の石版で再現された図譜には、アメリカ産の鳥がその生息する周辺環境とともに原寸大で再現されている。これらの図譜は一枚の絵の中に性別、年齢の違う個体を配し、さらに特徴的な部分が見える構図で描くことで識別図鑑としての機能をもつ。さらに生息環境や特徴的な行動を描いた生態図鑑でもある。背景となる植物や餌も学名とともに精密に示されている。合衆国のペリー提督はこの図譜一揃いを徳川将軍に贈ったと言われるが、その存在はいまだ確認されていない。

384　リンドウ科センブリ属の一種 (IMTB_HM0000383)

385　キンポウゲ科トリカブトの一種 (IMTB_HM0000384)

386　ジョン・ジェームス・ラフォレスト・オーデュボン
アオカケス／『アメリカ産鳥類図譜』(複製)／東京大学総合研究博物館研究部所蔵

387　セイタカダイオウ (タデ科) (IMTB_HM0000595)
2012年／ジャルジャレ・ヒマール (東ネパール)／乾燥標本／東京大学総合研究博物館資料部植物部門所蔵
ヒマラヤの高山帯に生えるタデ科の多年草。丈の低い高山植物が多い中、高さ1.5メートル、地際の直径は1メートルに達する。地面近くの葉は緑色をしているが、上部の葉 (苞葉) は乳白色をしており、光を半ば通すようになっている。

388　生薬標本コレクション
東京大学総合研究博物館資料部薬学部門所蔵
総合研究博物館では東京大学薬学部で研究対象とされてきたものをはじめ、1万5千種の生薬標本を所蔵している。戦前からの標本が多くあり、薬学分野での研究の変遷を窺い知ることができる。生薬とは植物、動物、鉱物など天然物に由来する薬用材料を医療や医薬品原料に供するものである。換言すれば、生薬を原料とする医薬品の成分を明らかにするには、歴史的な生薬サンプルが欠かせない。天然の資源が枯渇し、入手困難になるなか、学術研究機関に蓄積された生薬標本は新薬の製造開発の着想源にも、先端科学の研究用リソースとしても重要となる。

389　ジョン・ジェームス・ラフォレスト・オーデュボン
エボシクマゲラ、ラクーン・グレープ／『アメリカ産鳥類図譜』(複製)／東京大学総合研究博物館研究部所蔵

390　ジョン・ジェームス・ラフォレスト・オーデュボン
アメリカガラス、ブラック・ウォルナット／『アメリカ産鳥類図譜』(複製)／東京大学総合研究博物館研究部所蔵

391　ショクダイオオコンニャク (別名スマトラオオコンニャク) (IMTB_UT0000597)
2012年制作／2010年7月邑田仁撮影／インク・ジェットプリント／東京大学理学系研究科・理学部植物園にて開花／原寸高さ約1.5メートル／東京大学総合研究博物館研究部所蔵
インドネシア、スマトラ島のみに生息する希少種。小石川植物園では1991年に日本最初の開花を見たが、本写真の2010年の開花も日本で6例目と極めて珍しい。花が咲くと大きな仏炎苞が開き、その中から立ち上がっている棍棒状の花序付属体から強烈な腐臭を放ち、授粉する昆虫をひきつける。付属体の根元の仏炎苞の直径は1.3メートル、花の高さは3.1メートルにもなる。和名は全体の形をろうそくを立てた燭台に見立てたもの。

392　種子と果実コレクション
1950年代／黒沢幸子収集／乾燥標本／東京大学総合研究博物館資料部植物部門所蔵
種子は内部の幼い植物体を乾燥や水、捕食者他から守るため堅牢な構造をもち、自ら動くことのできない植物が親植物から離れて分布を広げていくためのいろいろな工夫をしている。果実はその中に含まれる種子をはじき飛ばす。種子や果実はコルク質を発達させて体重を軽くしたり、綿毛や翼他を発達させたりして、風や水流、海流によって遠くまで運ばれる。

393　熱帯の果実コレクション
大正から昭和初期／乾燥標本／東京大学総合研究博物館資料部植物部門所蔵
大正から昭和初期にかけて、東京大学植物学教室では、東南アジアやオセアニアにおける植物資源調査を活発に行った。ここに示す果実は、当時採集された標本の一部である。

394　タコノキ (タコノキ科) (IMTB_UT0000438)　1923 (大正12) 年／小笠原諸島
小笠原諸島に分布。和名は、茎の根元近くに出た根の様子が蛸のように見えることから。

395　ニッパヤシ (ヤシ科) (IMTB_UT0000434)　年代未詳／産地不明

代表的なマングローブ植物で、アジアとオセアニアの熱帯から亜熱帯に広く分布する。日本でも石垣島や西表島他に見られる。

396　ウミカラマツ類の一種
2011年購入／フィリピン産／乾燥標本／東京大学総合研究博物館研究部所蔵
樹木の「かたち」がそうであるように、海底での棲息環境への適応のため、全体が見事な扇形に生長している。微細な繊毛まで、枝分かれの原理が保持されている。欠損がほとんどなく得難い標本である。

397　サゴヤシ（ヤシ科）（IMTB_UT0000435）
1915年／トラック諸島（チューク諸島）
東南アジアからオセアニアにかけて分布し、広く栽培される。茎からでんぷんを取って食用とする。

398　ゴバンノアシ（サガリバナ科）（IMTB_UT0000437）
年代未詳／産地不明
琉球（西表島、石垣島）から台湾、東南アジア、オセアニアに広く分布。果実には毒があり、砕いて川に流して魚を捕る「毒流し漁」に用いる。和名は、果実の形が「碁盤の脚」に似ていることからいう。

399　オオフウチョウ（IMTAc_UT0000024）
20世紀後半／剝製標本／東京大学総合研究博物館研究部所蔵
ニューギニアの森林に分布する。フウチョウ科鳥類の雄は飾り羽を広げ、雌に対し求愛ダンスを行う。飾り羽の色や形は種によって様々である。フウチョウ類の羽毛は装飾として用いられたために乱獲されたが、現在は国際的に商取引が規制されている。

Greetings

The Intermediatheque is an open facility jointly operated by Japan Post and the University Museum, the University of Tokyo (UMUT), with the mission of contributing to society through its enlightenment and the diffusion of science.

In this facility, the scientific and cultural heritage accumulated by the University of Tokyo since its foundation in 1877 is permanently exhibited. The exhibition furniture mainly consists of items which were actually used for education and research. Most of it dating back to the Imperial University era, the solemn atmosphere it produces may lead some to feel as if they had time-slipped in the 19th Century. However, our intent is by no means nostalgic. Our aim is precisely to span a period of three centuries, from the Golden Age of Natural History in the 19th Century to the establishment of a highly developed information society in the 21st Century. To present an *outlook on the world*, of which the ages to come should not lose track, such is our aim.

Having been so far a storage space for our cultural heritage, museums have been solely considered as a place for exhibitions. However, by simply fulfilling this function, museums can no more satisfy the demands emanating from our 21st-Century society. Museums now have to perpetuate a comprehensive view of how we humans understand the world surrounding us. At the same time, they have to investigate on ways to draw new knowledge and means of expression from things and collections gathered there, and to present the fruits of such research. An experimental arena stimulating the dialogue of various means of expression through the practice of academic research and prompting the birth of new results, such is the Intermediatheque.

Historical scientific specimens accumulated in the University are most certainly a heritage from the past. However, at the same time, they constitute a resource which we should activate now while facing the future. In order to demonstrate this, we collected as much historical heritage as possible and modified its design for a new use, in accordance with our contemporary needs. In a society facing its limits in obtaining natural resources and energy supplies, it is no exaggeration to state that the task of *redesigning* accumulated extant objects is an urging issue confronting humankind. Aware of this current situation, we intend to pursue our activities by further aiming at the fusion of leading-edge technology and traditional crafting techniques. Our slogan, "Made in UMUT" (the University Museum, the University of Tokyo), thus contains a modest but strong message for the generations to come.

Our strong hope is that the Intermediatheque will be loved and supported by a wide audience in its further development.

Yoshiaki Nishino
Director

Scientific Specimens and Design
Yoshiaki Nishino

Isn't there something different about the things exhibited across the Intermediatheque, when compared to general museums? Some visitors may have this impression. It actually ought to be that way. You understand this immediately as you walk across the exhibition space: whether it be paintings or specimens, there is nothing commonly called *famous* to be found anywhere. What is more, the way of exhibiting appears to be different from what we usually see. That is because we did not adopt an evident way of exhibiting based on such criteria as chronology or classification, and from the outset, we did not have any intent to show the objects systematically.

Indeed, there may be visitors who feel confused at the Intermediatheque exhibitions. Some may even declare that not only there is no defined route in viewing them, but the space being totally unified, it is also difficult to distinguish between what is an exhibit, and what is not. In any case, it is clear that the Intermediatheque exhibitions do not follow ordinary exhibition methods. But why? It is partly because the Intermediatheque is different from general museums both in its formation and its social mission, as it is a joint research facility within a university, or in other words a university museum. But to most people, the term "university museum" may not provide a clear image. For the time being, please consider that since it was founded anew within the University of Tokyo in spring of 1996, the University Museum is both a research and educational institution promoting all leading academic disciplines, and an innovative and experimental museum questioning society about new curatorial methods and activities through projects such as Digital Museum, Mobile Museum or School Mobile.

Given these antecedents, if we apply the ordinary concept of museum to the Intermediatheque, many of its aspects remain incomprehensible. But please stop a moment and think. Currently in Japan, there are almost 6000 museums and equivalent or similar facilities, coexisting in a complete disorder. I won't mention here minor museums making an asset of their originality, or private museums dedicated to a particular theme. However, when we consider temporary exhibitions held in most national and public museums, although their title may differ, isn't the exhibition concept always similar? Isn't this recurrent similarity the current situation of museums? The reasons are easy to understand. Not only the museum professionals including curators, but also the public appreciating the fruits of curatorial activity, are bound by a set view on museums: they think that museums should be a certain way, with a certain type of exhibitions, etc. Here, I have an inquiry. Is it acceptable that exhibition installations look everywhere similar? Shouldn't we strain our brain so that each museum can produce an original world? Can visitors satisfy their inherent curiosity by simply walking along a predetermined exhibition route? It is now held that good museum exhibitions should be easy to understand, but if *understanding* is the supreme reward of a museum experience, what is there to be appreciated beyond? At the Intermediatheque, which proclaims to be an experimental facility, there are numerous issues we would like to raise.

Now, is the experience of visiting an exhibition space exclusively something to be understood intellectually? An intellectual understanding is a receptive attitude in which we attempt to understand what we have grasped through signs and words. Is this the only possible mode of reception in our appreciation of museums? This is where the Intermediatheque raises a new issue. Encountering exhibited things: this process starts with grasping the object with the eye, and continues with actually recognizing it by confronting it to memories accumulated in our brain. Grasping an object visually is mostly an act of reminiscence and of kinesthesia. Therefore, shouldn't there be exhibitions prompting us to an experience of working our sensitivity, of mobilizing as much as possible all of our senses?

If we decide from the outset that museums should be a certain way, exhibitions will become uniform and rapidly trite, and eventually tiresome. Museums, which should fundamentally be a place for the activation of our intellectual curiosity, will eventually deteriorate within the cultural ecosystem. We cannot but help thinking that such a tragi-comical situation is progressing among public cultural facilities. Isn't it necessary to reconsider the fact that a homogenized cultural environment is a particularly poor basis for creation? If we think of it, living organisms in the natural world are diverse, and it is precisely this diversity which attests to nature's abundance. Similarly, can't we state that it is by carrying out original exhibition principles in each museum with a competitive spirit that we will become culturally richer? If we are to take pride in a rich cultural matrix, first of all, don't we have to be able to clearly state to others our difference? University museums are a research and education facility attached to the university, and they are pilot museums. Therefore, we do not follow others, and neither do we duplicate without any reconsideration techniques which are already dated. That is precisely why we stick to our own method, and have kept on trying to refine it.

The Concentration of Things

As a result of education and research, a considerable amount of things were collected and produced in the University. Or course, this collection and production process is still pursued ceaselessly. Now, we resume this process under the single word of *collection*, but how was this actually carried out? In fact, ways of collecting within the University are diverse. Until recently, regulations concerning the protection of the environment and the flow of cultural heritage were not that strict. That is why researchers who travelled abroad were able to bring back research materials rather freely from every region of the world. If we put apart individual field research, the documents brought back by large scientific research groups organized by the University are considerable in volume. In the case of botanical specimens, hundreds or even thousands of which were collected every year, the collection kept on augmenting, and it was not rare to keep on describing and classifying its items ten or twenty years after returning from the expedition. Of course, this is not limited to the field of botany. It is the same for zoology, mineralogy or geology, as well as archaeology, anthropology or ethnology. Such longtime and steady work was deemed necessary to establish an important research corpus.

In the University, it is not rare to acquire research materials when necessary. Following the Meiji Restoration, the new government in its attempt to educate the forerunners of the modern Japanese State, even before regulating the education system, made the acquisition of a considerable amount of specimens and models from developed Western countries. Educate in the presence of real *things*: the pedagogic effects of such a system are as astonishing nowadays as they were in the past. For that purpose, the new Meiji government tried to gather the latest educational materials as rapidly as possible. This educational policy proved extremely sound. If we look back at this period, the judgment of the officials in charge of academia was effectively wise. The acquired specimens naturally served for research purposes, but they were first and foremost educational materials. First of all, show the real thing, or a sample, a model or a reproduction; by showing things, prompt a better understanding. Such things used in educational courses are usually named reference specimens. As they were collected, produced or bought with the aim of showing them to students, the *reference specimens* cover a wide range of representative things. Not only are they classified systematically, but their form and appearance are perfect. In comparison, what is the current situation? Education based on real things is obviously declining, and a virtualized form of education based on written information and digital images is dominant. Reference specimens are no longer used as traditional academic tools, and they are now regarded in research rooms and classrooms as cumbersome. It is a consolation in sadness that such specimens, instead of being discarded, were collected and reorganized by the University Museum. With their exhibition at the Intermediatheque, these salvaged scientific specimens now regain not only their true value as educational materials, but first and foremost their charm as things.

In addition, things and collections were collectively donated or entrusted to the University from outside parties. Naturally, the benefactors included many people related to the University, such as graduates and professors, as well as their descendants. In such cases, it is said that the scientific value of things depends heavily on their history and origin. However, the scientific value of things is only part of their overall value. That is why any hasty judgment is forbidden: in later times, the value of a thing unknown until then can be discovered. Also, it is certain that some things gain value as a certain period of time, say fifty or a hundred years, passes on. There are also cases in which a researcher in a given discipline rediscovers significant value in things which were discarded as waste in another discipline. If the University Museum positively accepts as many donations and entrustments as possible, it is because it has drawn the moral from these numerous precedents. Things do not hold an intrinsic, established value. When we discover in a thing a value unseen until then, we attribute a new value to it.

The Production of Things

The forms of production of things are also extremely diverse. At the University, activities such as the redaction of articles and the publication of research are conducted daily, and things have always remained for posterity as a product of that activity. Courses and practical exercises also lead to the production of things. In such sites where research is conducted, an experimental apparatus as well as auxiliary instruments are necessary for experiments and trials. Drawing and measuring instruments also become indispensable. To give another example from the Meiji era, it is particularly significant that among the heritage of the Imperial College of Engineering, which was later to become the School of Engineering, the oldest extant thing is a set of drawing and measuring instruments. It shows how the modernization of Japan began with the fortification of the foundations of scientific research. In Japan, the centennial planning of the modern State began with the acquisition of a scale.

It is said that research begins with things, and ends with them.

Whether the episode is actually true or not, just as Newton discovered the law of universal gravitation by watching an apple fall, the existence of things is always behind the birth of a new idea as a prerequisite. Once that idea is given form, it logically leads to the production of another thing. In the research rooms of Engineering or Medicine faculties, test pieces, samples and models are produced on a daily basis. We generally qualify the direct object of our research as primary materials, but those are not the only materials pertaining to education and research. As a result of research, secondary materials are produced, which in turn give birth to other subsidiary materials. Wherever research and education are conducted, materials thus multiply as in a chain reaction.

However, this is not the only reason why things are accumulated in the University. In places devoted to education and research, such as classrooms, laboratories and research rooms, minimum equipment such as tables, chairs and furniture are necessary, as well cabinets and bookshelves to store various research materials and documents. Such pieces of furniture have been individually numbered at the time of their acquisition, and in each faculty of the University, they have been registered as equipment. In such a public university as the University of Tokyo, a small label indicating the faculty, the registration number and date has been affixed on all acquired goods. When such labels have fortunately been preserved, it is possible to trace the origin of things. However, this doesn't mean that such registered items have all survived. In fact, once the initial cost of each equipment was amortized, it was customary to discard it after having removed its label. Of course, everything was not mechanically thrown away. *What a waste...* Such a natural feeling arouses in the heart of anyone, in any given period. Nonetheless, under the pretext of renewing the equipment, there are many cases in which precious things have been discarded.

Such things include official documents related to the general management of the University, commemorative albums, portraits and busts of professors, as well as commemorative goods, medals, uniforms, and memorials, all things and records pertaining both to the official history of the University and the teachers' personal history. They form what we usually name the historical archive of the University. Records of the successive presidents of the University, starting with the personal documents of Hiroyuki Kato, the first president of the Imperial University, also belong to this category. Such corpuses bear the denomination of *library*, *collection* or *historical document*. They are preserved in a single place, without being dispersed. Such documents as the collection of architectural photographs of the University campus taken by the 14th president Yoshikazu Uchida after the Great Kanto Earthquake, or the architectural plans of the campus buildings left in the headquarters of the University, are fundamental in that they are deeply related to the existence of the University, the transformation of its facilities as well as its past accomplishments.

The Utilization of Things

Through its educational and research activities, the University thus has given birth to a wide diversity of things. At the University Museum, all these things are customarily called *scientific specimens*. Of course, scientific specimens are extremely diverse, in their composition, their dimension and their form. However, they all hold a common characteristic: they all reflect the core of education and research activities in their diversity. Moreover, with the exception of donations and acquired equipment, as most scientific specimens form a corpus collected and constituted by teachers and researchers themselves, their origin and history are certain, and resist any scientific verification. From this point of view, University Museums differ from general museums, which have to rely upon the description of a third party to establish the origin and provenance of their collection. If we meticulously examine a given set of scientific specimens, we can have a clear view of the working methods of the researcher who produced them. By glancing over the scientific specimens left by their predecessors, researchers who carry our future can thus verify their actual position by having an global and diachronic understanding of it.

In the case of universities specialized in a given discipline such as law, engineering, medicine, science or agriculture, most scientific specimens consist of documents and records pertaining to research or to professors of that given specialized discipline. In the case of general universities, the accumulated scientific specimens cover a wide range of disciplines. This is why in the case of the University of Tokyo, which boasts the longest academic history in Japan, the diversity and overall quantity of accumulated specimens reach a scale unseen elsewhere.

How many specimens were actually accumulated at the University of Tokyo? To be frank, no one really has a precise account of the total quantity. In fact, even now, specimens are continuously being produced. To give some data which is beginning to date, according to statistics taken when the University Documentation Center was reformed and expanded to form the University Museum in 1996, about 6 million specimens were counted in the whole University, 2.4 millions of which were stored within the University Museum. Obviously, some precisions are necessary when giving such statistics, as they depend on how we actually count things as *one set* of specimens. At the opening of the Intermediatheque, the exhibits amounted to about 7000 items in totality, which formed an estimated 1000 sets. In other words, given statistics do not always reflect reality. Moreover, as the number above only covers registered items, it is difficult to imagine the total number of items if we are to include those which are not registered yet. At the University Museum since its renewal, there are some 50 000 to 100 000 sets of specimens which are added to the collection yearly. It can thus be estimated that its collection now easily surpasses 4 million sets in total.

From the University of Tokyo collection, considerable in its overall volume, we selected scientific specimens remarkable for their rarity and scientific value, and exhibited them at the Intermediatheque. The exhibits cover a wide range of disciplines, from cultural history to natural history, including engineering and the history of science. The collections are equally diverse from a spatial and temporal point of view. Materials pertaining to planetary science, such as stardust or meteorites, belong to a cosmic timescale. Petrology and mineralogy specimens related to the constitution of Earth are measured on a geological timescale. Fossilized human bones and stone implements in prehistoric archaeology are measured in hundreds of thousands to millions of years, whereas the heritage of ancient civilizations in thousands of years, and artistic and cultural heritage in centuries – a distribution in time which can be grasped on a human scale. Of course, these items are widely distributed in space, to cover all regions of Earth.

The University of Tokyo collection consists of collections ranging over a wide diversity of disciplines. It can thus be considered as an aggregation (or collection) of collections. But there is more to it. Each collection, when compared to similar ones in research institutions in Japan and abroad, can be said to be first-rate. Since its foundation in 1877, the University of Tokyo has been leading education and research in Japan as a public university. When taking this fact into account, it is no surprise that the collections accumulated within the University are outstanding. To put it in an extreme way, from a domestic point of view, the University of Tokyo collection concentrates many illustrious items in each discipline: the first discoveries made in a given field, the oldest extant historical documents, the biggest specimens, the first samples ever made, things particularly remarkable for their rarity and scientific value. It is part of this collection which adorns the Intermediatheque exhibition space.

The Presentation of Things

The items stored within the University Museum thus constitute a collection of *scientific specimens* grouped in one single place, according to the process described above. This is why, when organizing exhibitions, we cannot adopt a policy often seen in general museums and valuing masterpieces above all. However, the fact that we do not possess masterpieces does not imply that we have to limit our ambitions. Accepting the fact that even if wanted to, we could not depend on treasures or famous pieces, we had to ask ourselves what we could actually do, given these conditions. Why not then consider the given situation, which obliges us to give birth to new and creative ideas, as a productive factor for the realization of exhibition design never seen before? The conception of the Intermediatheque originates from such a positive and constructive thought. To give just an example, in the case of a Natural History exhibition, we can either align large quantities of specimens of differing species to demonstrate the wide diversity of species, or we can chose to show many specimens of a given species to illustrate its wide distribution. Moreover, scientific specimens produced in the process of university education and research are filled with elements which pertain to the eccentricity and interest of the scientists' thoughts, which originally led them to show interest in such things. This is probably the reason why such scientific specimens also appear attractive to the eye of the amateur.

How should we consider our encounter with things exhibited in museums? The exhibitions at the Intermediatheque address such a fundamental issue. As far as the exhibited items were initially collected for research and education purposes, they are not necessarily suitable for an aesthetic appreciation. Given their respective quality, these scientific specimens are interesting from a visual point of view when considered globally as a collection. This is why we have to think out a way of showing them, different from the ordinary methods.

This original conception of exhibition design includes a wide array of elements, from strategies in visual psychology and spectacular stagecraft effects, to creative ideas inspired by the visual arts and a sense of composition based on museum technology. How to stimulate creation and originality, and above all, considering the exhibits as objects of aesthetic appreciation, how to guarantee their visual beauty and the pleasure we take in their encounter: it is from such a standpoint that the exhibitions were conceived, addressing those issues from every angle. On the basis of such considerations, the Intermediatheque is interspersed with numerous pieces of furniture long used within the University campus, as well as old architectural elements. These have been modified so as to function within an exhibition space. There is also some exhibition furniture which has been made of discarded material collected during the destruction of buildings. At the Intermediatheque, we name such modified uses, transformations and assemblages ReDESIGN+, and we implement it positively. The fundamental policy of the Intermediatheque is to limit the consumption of resources and to reduce our impact on the environment by recycling things which have become useless.

Those visitors who have already seen the Intermediatheque have certainly noticed that there is no determined route in viewing the exhibition. Just like the researchers in natural history during their fieldwork who look around for animals and plants, the Intermediatheque visitors, instead of following a given route, can enjoy their encounter with things by wandering freely in the exhibition space according to their interest. The Intermediatheque was conceived to be such a place.

General museums have the status of educational facilities. This is why every effort is made to optimize their pedagogic impact through the selection and the order of presentation of the exhibits, as well as the contents and format of the explicative panels. Sometimes, in an excessive manner. At other times, in such proportions that it can be qualified as insensitive. Practically, this is done by providing explanations as considerate as possible, printed in big letters so as to be clearly visible. Nowadays, digital terminals are fixed close to the exhibits so as to provide additional data, and audio guides prompt a certain visiting route. Such exhibitions are seen here and there, and these technologies are certainly propagating within the museum world. However, we have never heard that the actual effectiveness of such exhibitions in teaching and transmitting information was ever verified. I do not intend to discuss here the legitimacy of such technologies. However, at the Intermediatheque, according to our policy and method of privileging an environment for aesthetic appreciation, we have ensured that verbal information such as explanation panels do not create any visual impurities.

All in all, where is the educative aspect of putting up, without any hesitation, insensitive exhibitions which shamelessly expose badly designed panels and illustration boards? Such a sight is desperately common in museums throughout Japan. At the Intermediatheque, we directly address the *acultural* aspect of such public educational methods. In doing so, we do not in the least disregard the transmission of knowledge. However, it is clear that aesthetically poor exhibition design has no redeeming features. To improve this, it suffices to pay attention to design. Isn't it by exhibiting charming things in a beautiful way that we can contribute to the cultivation of our sensibility? The Intermediatheque is not a museum which is to be conceptually understood through reading, but one which stimulates creativity through the act of seeing.

005-006 *Tomistoma machikanense* (*Toyotamaphimeia machikanense*), Crocodilians (IMTAb_UT0000129)
Pleistocene layer / Resin (replica of skeleton specimen) / Donated by Harano Agricultural Museum / UMUT
The original fossil was found in a pleistocene layer (0.3-0.5million year) in Machikane hill, Osaka, Japan in 1964. This crocodilian was given its name from the place where the original fossil was found.
007 "Egg in Space" Installation
2007 / Yoshiaki Nishino + Sergio Calatroni + Hiroyuki Sekioka + Hiroto Nakatsubo / *Aepyornis* egg (replica made by Kyoto Kagaku Co.), marble, brass, cypress base / UMUT
This installation offers an interpretative counterpoint to Constantin Brancusi's "Bird in Space." While producing the series of works revolving around the theme of the bird, Brancusi also paid attention to the form of the egg. The work in which he placed an egg on a metallic plate is well known, but he never actually erected an egg. In making this installation, materials which suit the sculptor best were chosen for the base.
008 Electrical Engineering Instruments (IMTF0000297)
Date unknown / Metal / Formerly a collection of Department of Electrical Engineering, the University of Tokyo / UMUT
009 Engineering Instruments
010 Turridae (Mollusca: Gastropoda)
1990's / Amami Oshima, Kagoshima / Dried specimen / UMUT
011 Kazumi Taguchi image (IMTE_MD0000198)
1894 / Painted by Katsuzo Takahashi / Oil on fabric / Department of Anatomy, The University of Tokyo
Portrait of the oldest in the University of Tokyo. Kazumi Taguchi(1839-1904) first professor of the Department of Anatomy, University of Tokyo. Anatomical Society of Japan has been established in the year 1893, he served as its President. Portrait is of this time. Painter of the author Katsuzo Takahashi(1860-1917) year moved to the U.S. 1885, he studied painting at the California School of Design in San Francisco. Returned in 1893, I opened a laboratory Shibayama in Shiba.
012 Steam Engine Model (IMTF0000036)
1870's / Elliott Brothers Ltd. / London (England) / UMUT
Educational material used to popularize the structure of power transmission.
013 Ujihiro Okuma (IMTJ_UT0000186)
Statue of Moses

1879 / Plaster / Formerly a collection of the Technical College Art School / UMUT
Ujihiro Okuma (1856-1924) entered the Technical College Art School at the time of its foundation, and studied sculpture under Ragusa. Two plaster portraits made while he was attending school still exist. This work is one of them, and bears the mention "Completed in February, Meiji 12 *michelan* Okuma Ujihiro." It is thought to be an imitation of Michelangelo's *Moses*.

014 Bust of Charles Dickinson West (IMTJ_UT0000185)
Date unknown / Sculpted by Kazumasa Numata / Resin / Formerly a collection of the Department of Industrial Mechanical Engineering, The University of Tokyo / UMUT
Charles Dickinson West (1847-1908) was an Irish teacher employed by the School of Engineering. Born in Dublin, he studied Mechanical Engineering at the Trinity College at the University of Dublin, and graduated in 1869. He then worked at the Bergenhead Steel Company for five years, where he acquired knowledge on ship building. He arrived in Japan in 1882 to replace Henry Dyer at the School of Engineering, where taught Mechanical as well as Marine Engineering. He remained in Japan until his death in 1908.

015 Cattle Horn (*Bos primigenius*)
Date unknown / Dried specimen / Private collection
The horn consists of core living bone and surrounding sheath. This specimen is sheath detached from the skull. The horn is a pointed structure on the head which continuously grows throughout the animal's life. On the other hand, the antler of the deer is a one-piece structure which sheds every year. The outer layer if the horn consists of Keratin, just as human nail. This specimen is polished so showing the extending and twisting growth process.

020-021 Large skeleton specimen

022 Crustacea Dried Specimen Collection
1880's / Isao Iijima (?) / Formerly a collection of the Department of Zoology, the University of Tokyo / UMUT

023 Collection of Reptiles and Amphibians
1880's / Chujiro Sasaki (?) / Formerly a collection of the Department of Zoology, the University of Tokyo / UMUT

024 Soft-shelled turtle (IMTAd_UT0000023)
Date unknown / Dried specimen / UMUT

025 Frog Skeleton Specimen Collection (IMTAd_UT0000001-13)
1880's / Chujiro Sasaki (?) / Formerly a collection of the Department of Zoology, the University of Tokyo / UMUT
Behind their curious appearance, frogs hide a peculiar skeleton allowing for their characteristic jumping locomotion. It is such a structure, especially the part from the hip to the limbs, similar to no other animals, which generates that spectacular thrust. In spite of such an adaptation, the scull and ribs are rather primitive.

026-027 *Aepyornis* Egg (replica)
History unknown / Replica made in 2012 / Color on resin / Donated by Natural History Museum, Aix-en-Provence (France) / UMUT
Among animal species known to us, the *Aepyornis* is the one which lays the biggest egg. The egg's volume is 7 to 9 liters, which is the equivalent of about 180 hen eggs. The egg is representative of a perfect form issued from the natural world, and as it contains life within its shell, it can also be seen as a cosmic globe encompassing all beings.

028 Collection of Mammal skeleton specimen

030 Japanese weasel (male) (IMTAb_UT0000041)
2005/ Skeleton specimens / UMUT
Eurasian badger (IMTAb_UT0000043) 2003 / Skeleton specimens / UMUT

031 Red fox (IMTAb_UT0000001)
Date unknown / Skeleton specimens / UMUT
Eurasian badger (IMTE_MD0000028)
Date Unknown / Skeleton specimens / Formerly a collection of Department of Medicine, UMUT / UMUT
Lutrinae, a species of Otter (IMTE_MD0000021)
Date Unknown / Skeleton specimens / Formerly a collection of Department of Medicine, UMUT / UMUT

032 Reeves's muntjac (female) (IMTAb_UT0000003)

1957 death / Skeleton specimens / UMUT
Javan gold-spotted mongoose (IMTAb_UT0000040)
2001 / Amami Island, Kagoshima / Skeleton specimens / UMUT
Raccoon dog (IMTE_MD0000031)
Date Unknown / Skeleton specimens / Formerly a collection of Department of Medicine, UMUT / UMUT
European rabbit (IMTE_MD0000026)
Date Unknown / Skeleton specimens / Formerly a collection of Department of Medicine, UMUT / UMUT
Hedgehog (IMTE_MD0000027)
Date unknown / Skeleton specimen / Formerly a collection of Department of Medicine, UMUT / UMUT

033 Hedgehog (IMTE_MD0000027)
Date unknown / Skeleton specimen / Formerly a collection of Department of Medicine, UMUT / UMUT
Sika Deer (Embryo) (IMTAb_UT0000057) 2003 / Skeleton specimen / UMUT

034-035 Collection of Mammal skeleton specimen

036 Cat (IMTAb_UT0000039) Date unknown / Skeleton specimen / UMUT
Polecat (IMTE_MD0000029)
Date unknown / Skeleton specimen / Formerly a collection of Department of Medicine, UMUT / UMUT

037 Weasel (IMTE_MD0000032)
Date Unknown / Skeleton specimens / Formerly a collection of Department of Medicine, UMUT / UMUT
A species of Snake (IMTE_MD0000058)
Date Unknown / Skeleton specimens / Formerly a collection of Department of Medicine, UMUT / UMUT

038 Red Deer (Male) (IMTAb_UT0000064)
2003 / Mongolia / Skeleton specimen / UMUT
This species is a large deer that distributes in Europe, Central Asia and a part of Africa. The body height of the male can reach 150cm and weigh 200kg. The antler, which is the deer's specific characteristic, is a bony structure that grows in spring and sheds annually after the copulating season in autumn. Big antlers are an appeal to females, as well as a symbol of the deer's status toward other males, and can also be used for fighting. This red deer specimen was acquired during overseas research activity conducted by the UMUT staff.

039 Ostrich (IMTAb_UT0000150)
2005 (?) / Skeleton specimen / Formerly Inokashira Park Zoo collection / UMUT
The heaviest among living birds, the ostrich weighs about 130 kilograms. It has two toes on each leg (whereas birds usually have four), supposedly an adaptation to running on flat terrains. If the ostrich cannot fly, however the skeleton of its wing is not completely degenerated and it is used for keeping balance while maneuvering or for displaying behavior. The bird can run at a speed of 70km/h to escape from large predators.

040 Figurative Form
2011 / Toshimasa Kikuchi / Cypress / Private collection
From olden times, Japan has had a tradition of wooden skeletons. Until the late Edo period, due to Buddhism restriction, using actual skeletons in medical research facilities are forbidden. Such facilities would ask a sculptor who specialized in creating Buddhist statues to make a wooden skeleton for them. Modern sculptors are attempting to revise the tradition.
The first *Ardipithecus ramidus* fossil
4.4 million years ago / Discovered in December 17, 1992 / Ethiopia / Research cast
Ardipithecus ramidus was discovered in 1992, as the oldest known human ancestor at that time. In 2009, a partial skeleton and other fossils revealed that *Ar. ramidus* represents a newly recognized primitive evolutionary grade ancestral to humans. *Ar. ramidus* possessed a grasping foot as in apes, but walked upright like humans. It appears to retain characteristics close after the human-chimpanzee split. The exhibited specimen is an upper right third molar, the first *Ar. ramidus* fossil found at Aramis, Afar Rift, Ethiopia.

041 The discovery of special tombs associated with gold ornaments in Kuntur Wasi site, Peru ca. 800 B.C.

Kuntur Wasi is located in the northern Peru. This site functioned as a ceremonial center during the late Initial Period and the Early Horizon (ca. 1000-250 B.C.). The Archaeological Mission of the University of Tokyo launched excavations at this site in 1988. In 1989 and 1990 field seasons, four tombs were discovered in association with many gold ornaments as offerings. They are the oldest gold artifacts that were metallurgically refined in the Americas. The pieces exhibited here are the replicas of the gold artifacts from three tombs and made of the alloy of the same material composition as their originals.

Gold Crown Decorated with Fourteen Human Faces
Gold Ear Ornaments representing Profiles of Jaguars
Gold Ear Spools
Gold Crown Decorated with Five Jaguar Faces
Gold Mouth Mask representing a Jaguar and Twins
Gold Mouth Ornament representing a Jaguar Face with a Snake-coiled Eye and a Quadrangular Eye

042 Bottlenose Dolphin (IMTE_MD0000035)
Sep. 1957 / Skeleton specimen / Formerly a collection of Department of Medicine, UMUT / UMUT

043 Japanese Seabass (IMTE_MD0000011)
Date unknown / Skeleton specimen / Formerly a collection of Department of Medicine, UMUT / UMUT

044 Acropomatidae, a Species of Lanternbellies (IMTE_MD0000019)
Date unknown / Skeleton specimen / Formerly a collection of Department of Medicine, UMUT / UMUT

045 *Carassius* sp., a Species of Crusian Carp (IMTE_MD0000017)
Date unknown / Skeleton specimen / Formerly a collection of Department of Medicine, UMUT / UMUT

046 Sea Robin (*Chelidonichthys spinosus*) (IMTAe_UT0000002)
2012 / Skeleton specimen / UMUT

047 A species of Moonfish (*Mene maculate*) (IMTAe_UT0000003)
2012 / Skeleton specimen / UMUT

048-049 Minke Whale (IMTAb_UT0000126)
Date unknown / Skeleton specimen / Formerly a collection of the Faculty of Agriculture, the University of Tokyo / UMUT
The species is the second smallest among baleen whales (Mysticeti), although it grows up to 7-8m long and weighs about 8t. As a result of their adaptation to aquatic environment, their hind limbs have completely disappeared, and their spine liberated from gravity has a simple structure. Their scull shows signs of specification too: their nostril has changed to a spiracle (blow hole) opening upper side of the head. Instead of teeth, they have a baleen system on the upper jaw which filters small fishes or krill after swallowing them with along water. The whale's visual sense is not highly developed, and it mainly resorts to auditory signals and vocal communications.

050 Japanese Macaque (IMTAb_UT0000171) 2006 / Skeleton specimen / UMUT

051 Collection of Bird Skull
Date unknown / Formerly a collection of Department of Medicine, UMUT / UMUT
Birds have lost their teeth in the way of their evolution. Thereafter, the form of their bills is closely related to their food habits. Every specific *form* of their bills is adapted to their specialized foraging behavior, such as tearing preys, catching fishes, cracking seeds or picking up small animals in mud.

052 Collection of Mammal Skull
Date unknown / Formerly a collection of Department of Medicine, UMUT / UMUT
Mammals developed heterodont dentition, which is a functional differentiation in teeth morphology into incisors, canines and molars. Their teeth thus clearly indicated their food habit. Adding to it, the sculls show species-specific morphological characteristics in the arrangement of jaw joints and muscles, the holding of sensory organs, brain protection as well as weight reduction. These specimens embody the *forms* necessary for survival, and the forms resulting from the long history of evolution.

053 Sturnidae (A Species of Starling) (IMTAc_UT0000026)
Skeleton specimen / 1960s'

054 Japanese Macaque 2002 / Skeleton specimen / UMUT

055 Maroon Langur (IMTE_MD0000169)
Date Unknown / Skeleton specimens / Borneo / Formerly a collection of Department of Medicine, UMUT / UMUT

Celebes Crested Macaque Or Moor Macaque (IMTE_MD0000170)
Date unknown / Skeleton specimen / Formerly a collection of Department of Medicine, UMUT / UMUT

Eastern Grey Kangaroo (IMTAb_UT0000182)

056 Orangutan (IMTE_MD0000171)
Date Unknown / Skeleton specimens / Borneo / Formerly a collection of Department of Medicine, UMUT / UMUT

057 *Pelecanus* sp., a species of Pelican (IMTE_MD0000172)
Date Unknown / Skeleton specimens / Formerly a collection of Department of Medicine, UMUT / UMUT

058 *Macropus*, a Species of Kangaroo (IMTE_MD0000034)
1910 / Skeleton specimen / Formerly a collection of Department of Medicine, UMUT / UMUT

Human (IMTE_KO0000001)
Date unknown / Skeleton specimen (replica) / UMUT

059 Right Humerus of Goat (*Capra hircus*)
2002 / Ogasawara Village, Chichi-jima (Tokyo) / Private collection
Most animals use their limbs for locomotion. The bone strength, the joint position and its mobility, as well as the muscle origin/insertion and orientation all depend on the shape of bones. The locomotion ability of the animal therefore varies according to the shape of its bones. Locomotion of cursorial animals such as the goat (Bovidae, Cetartiodactyla) relies on the extension/flexion of the joints on the sagittal plane.

060-062 Collection of Avian skeleton

063 Sparrow Hawk 2011 / Skeleton specimen / Private collection
Eurasian Oystercatcher (IMTE_MD0000051)
Date Unknown / Skeleton specimens / Formerly a collection of Department of Medicine, UMUT / UMUT

064 *Corvus* sp., a species of Crows (IMTE_MD0000054)
Date Unknown / Skeleton specimens / Formerly a collection of Department of Medicine, UMUT / UMUT

065 *Vanellus* sp., a species of Lapwings (IMTE_MD0000053)
Date Unknown / Skeleton specimens / Formerly a collection of Department of Medicine, UMUT / UMUT

067 A species of Parakeet (*Aratinga* sp.)
2010 / Skeleton specimen / Private collection

068 Herring Gull 2012 / Skeleton specimen / Private collection
Pied Avocet (IMTE_MD0000048)
Date Unknown / Skeleton specimens / Formerly a collection of Department of Medicine, UMUT / UMUT

069 Domestic Cock (IMTAc_YZ0000003)
Date unknown / Skeleton specimen / Formerly a collection of Yaizu Fishery High School, Shizuoka / UMUT

070 Collection of Avian and Bat skeleton

071 *Eudocimus*, a species of White Ibis (IMTE_MD0000049)
Date Unknown / Skeleton specimens / Formerly a collection of Department of Medicine, UMUT / UMUT

Copper Pheasant (IMTAc_UT0000013) Date unknown / Skeleton specimen / UMUT

072 Jungle Crow (IMTAc_UT0000015) 2003/ Skeleton specimen / UMUT

073 Ryukyu Flying Fox (IMTAb_UT0000164) 2004 / Skeleton specimen / UMUT

074 Horus Statue
Date unknown / Bronze / Department of Pharmaceutical Science, UMUT
Horus is the god of the sky and the sun in ancient Egypt. He is usually depicted as a bird of prey, and this sculpture is a falcon. Horus's right eye represents the sun and the left one the moon. It is said that the shape of the left eye looks like the letters of alphabet, "Rx", and this mark has been used as a sign in prescriptions. From this, it can be considered that this sculpture had decorated the buildings of the Faculty

of Pharmacology in the University of Tokyo as a motif deeply associated with pharmacology.

075 Deer skulls

076 Limbs of large mammals

077 Skeleton of herbivorous mammals

078 Skull of herbivorous mammals

079 Cattle (IMTAb_UT0000072) 2003 / Skeleton specimen / UMUT
Pig (IMTAb_UT0000014) (IMTAb_UT0000015)
2003 / Skeleton specimen / UMUT

080 Skull of Goat and Sheep

081 Skull of Goat, Sheep and Gazelle

082 Goat (IMTAb_UT0000075) 1999 / Skeleton specimen / UMUT
Goat (*Sibayagi*, a Japanese var.) (IMTAb_UT0000077)
2002 / Skeleton specimen / UMUT
Mongolian Gazelle (Male) (IMTAb_UT0000076)
2004 / Skeleton specimen / UMUT

083 Skeleton of Turtles

084 Soft-Shelled Turtle (IMTE_MD0000041)
Date unknown / Skeleton specimen / Formerly a collection of Department of Medicine, UMUT / UMUT

085 A Species of Alligator (IMTE_MD0000059)
Date unknown / Skeleton specimen / Formerly a collection of Department of Medicine, UMUT / UMUT

086-087 False Killer Whale (IMTAb_UT0000127)
2008 / Skeleton specimen / Kept in Yokohama Hakkeijima Sea Paradise / UMUT
Unlike other toothed whales (odontoceti), this species as well as the killer whale attack large preys such as tuna or other whales. Not only do their well-developed teeth catch and keep small animals, but they can also cut out a chunk of meat from large preys. This species grows up to 6m long at the largest. The concaved structure on the forehead is a cavity containing a unique organ called melon. This organ is specific to toothed whales, and is dedicated to echo location, emitting high frequency sounds to detect preys.

088-089 Least Horseshoe Bat (IMTAb_UT0000038) 2005 / UMUT
Greater Horseshoe Bat (Male) (IMTAb_UT0000044) 2003/ Fukuoka / UMUT
Pteropodidae, a Species of Megabats (IMTE_MD0000047)
Date unknown / Department of Medicine, UMUT
A species of Bat (IMTAb_UT0000166) 1968 / Stuffed specimen / UMUT

090-091 Small Animal Skelton Specimen Collection UMUT

092 Japanese Hare (Female) (IMTAb_UT0000042)
2004/ Tokyo / Skeleton specimen / UMUT
Japanese Lesser Flying Squirrel (IMTAb_UT0000036) 2006
Japanese Toad (IMTAd_UT0000019) 2006

093 European Mole (IMTE_MD0000025)
Date unknown / Department of Medicine, UMUT

094-095 Beetle Specimen Collection from France
Late 19th Century to 1970's / Purchased in 2010 / Dried specimens / UMUT

096 Beetle Specimen Collection from France
Late 19th Century to 1970's / Purchased in 2010 / Dried specimens / UMUT

097 *Cymbium amphora*
Date unknown / Locality unknown / Dried specimen / UMUT
This shell is considerably swollen and attains more than 30 cm in height. The animal is large, covering most of the shell and lacking an operculum. This species is carnivorous and deposits eggs in egg capsules after internal fertilization. Its habitat consists in shallow sandy bottoms in the tropical Western Pacific. The scientific name indicates a melon-shaped shell with an apical depression.
Molluscan Shell Specimen Collection
1880's / Formerly a collection of the Department of Zoology, the University of Tokyo / UMUT
Representative specimens of molluscan shells collected by the Zoological Institute of the former Tokyo Imperial University. The institute started collecting various animals in the 1870's, the shell collection constituting the largest part. Specimens from all regions

of the world were obtained through exchange with foreign institutes. Numerous tropical shells were also collected from Pacific islands by university staff. Conservation cases are made of paper-thin sheet of wood, with Japanese paper covering the wood and Japanese lacquer painted on the surface. This type of cases was most popular in Tokyo Imperial University.

098-099 Molluscan Shell Specimen Collection
1880's / Formerly a collection of the Department of Zoology, the University of Tokyo / UMUT

100-101 Masai Giraffe (Male) (IMTAb_UT0000150)
2009 / Skeleton specimen / Kept in Kobe Oji Zoo (named Shinpei, died age 8) / UMUT
Despite its astonishingly long neck, the giraffe only has seven vertebrae, like the neck of other mammals. If the giraffe's curious proportions have raised many discussions, the survival value of its long neck remains uncertain. Not only can they forage leaves on high trees, but male giraffes also fight by wreathing around with their necks. The giraffe also has considerably long legs, making it the highest animal in the world. On the other hand, the form of its trunk is common among herbivorous mammals.

104 Water Monitor (IMTAd_UT0000018)
Date unknown / Stuffed specimen / UMUT
This carnivorous species distributes in a wide area of South-eastern Asia, prefers the waterside and often swims. It can measure up to 2.3m.

Charonia tritonis
Date unknown / Okinawa (?) / Dried specimen / UMUT
The large-sized shell reaches more than 40 cm in length, but large specimens are difficult to obtain recently. The animal is carnivorous and preferably preys on starfishes. This species is particularly well-known as a predator of crown-of-thorns starfish (*Acanthaster planci*) which has a frequently destructive effect on corals. Its habitat is south to Kii Peninsula, mainly from Amami Islands to the tropical Pacific, in shallow-water coral-reef environments.

105 The University of Tokyo Mammal Stuffed Specimen Collection
The University of Tokyo collected countless of specimen in the long history of 130 years. In the days that image technology was less advanced, stuffed specimens had numerous values by showing real form of the strange animals. These have value still now, because some of them are rare nowadays and some tells our history. For example, earless seal was common in Hokkaido in past days, and Leopard Cat from Taiwan was "Japanese animal" in those days.

Oita Wild Bird Museum Mammal Stuffed Specimen Collection
The collection donated from Wild Bird Museum (Gifu prefecture, closed in 2008) by Keikichi and Masao Oita. The museum was established for education and reservation of wild birds, but stored some mammal specimen adding to a large collection of birds because the museum was to show the natural history of the country, also. Here we display these mammals, or our "neighbors", that live around us but difficult to see.

106 Small Indian civet (IMTAb_UT0000004)
Purchased 1976 / Stuffed specimen / UMUT

107 Japanese Hare (IMTAb_OT0000140)
1971 Nov. 19 / Formerly Oita Wild Bird Museum Collection / UMUT

Japanese Hare (IMTAb_OT 0000143)
Date unknown / Formerly Oita Wild Bird Museum Collection / UMUT

Japanese Hare (IMTAb_OT 0000144)
Date unknown / Formerly Oita Wild Bird Museum Collection / UMUT

108 Chinese leopard cat (IMTAb_UT0000005)
Date unknown/ Stuffed specimen / UMUT

Phocidae, a species of earless seal (juvenile) (IMTAb_UT0000002)
Date unknown / Stuffed specimen / UMUT

109 Weasel (IMTAb_UT0000007)
Date unknown/ Stuffed specimen / UMUT

Japanese red fox (IMTAb_OT0000146)
Date unknown/ Stuffed specimen / Oita collection, UMUT

Japanese raccoon dog (IMTAb_OT0000142)
1968 / Stuffed specimen / Oita collection, UMUT

110 A species of armadillo (IMTAb_UT0000006)

Date unknown/ Stuffed specimen / UMUT
111 Asian Black Bear (Juvenile) (IMTAb_OT 0000149)
Date unknown / Formerly Oita Wild Bird Museum Collection / UMUT
112 Japanese Water Shrew (IMTAb_OT 0000134)
1965 Aug. 6 / Formerly Oita Wild Bird Museum Collection / UMUT
113 Japanese Squirrel (IMTAb_OT 0000135)
Date unknown / Formerly Oita Wild Bird Museum Collection / UMUT
114 Japanese Weasel (Male) (IMTAb_OT 0000121)
Collected in 1974 / Formerly Oita Wild Bird Museum Collection / UMUT
115 Japanese Squirrel (Juvenile) (IMTAb_OT 0000137)
Date unknown / Formerly Oita Wild Bird Museum Collection / UMUT
116 Stoat (IMTAb_OT 0000117)
Date unknown / Formerly Oita Wild Bird Museum Collection / UMUT
117 Hedgehog (IMTAb_UT0000008) Date unknown / UMUT
118 Japanese Dwarf Flying Squirrel (IMTAb_OT 0000139)
Date unknown / Formerly Oita Wild Bird Museum Collection / UMUT
119 Cattle anatomy model (IMTAb_UT0000065)
Date unknown / Yamakoshi educational equipments and specimen Co. Ltd/ Tokyo / color on paper / UMUT
120-121 Small-Scale Models of Domesticated Cattle (IMTAb_UT0000066, 157-160)
1883-1892 / Max Landsberg in Berlin (Germany) / Color on plaster / Formerly a collection of the Department of Zootomy, Tokyo Imperial University and of the Department of Zootomy (Specimen Room), The University of Tokyo / UMUT
Plaster models of cattle reduced on a 1:6 scale. The models are scale-downed from existing individuals domesticated in The University of Veterinary Medicine in Berlin. Because the education of zoology and animal husbandry in Japan was influenced by Germany, these high-quality animal models for education were introduced from Berlin in the Meiji Era. The three breeds of Jersey, Shorthorn and Simmenthal can be morphologically compared.
122-124 Maritime Invertebrates
125 *Strombus latissimus*
Date unknown / Locality unknown / Dried specimen / UMUT
126 *Hemifusus tuba* Date unknown / Japan / Dried specimen / UMUT
Tectus niloticus Date unknown / Locality unknown / Dried specimen / UMUT
Hippopus hippopus
Date unknown / Iriomote Island, Okinawa / Dried specimen / UMUT
127 Busycon canaliculatum
Date unknown / Locality unknown / Dried specimen / UMUT
Cassis madagascariensis
Date unknown / Locality unknown / Dried specimen / UMUT
Rapana venosa Date unknown / Japan / Dried specimen / UMUT
Chicoreus ramosus
Date unknown / Locality unknown / Dried specimen / UMUT
Tridacna squamosa
Date unknown / Locality unknown / Dried specimen / UMUT
128-129 Maritime Invertebrates
130 Sea Urchin Specimen Collection
Sea urchins are invertebrate animals belonging to the class Echinoidea in the phylum Echinodermata. The animal is encased in a calcareous test which is globular, oval or discoidal in shape. The test is covered with spines of various length, which fall off after the animal's death. Sea urchins have adapted to a wide range of marine habitats from shallow to deep sea. The number of known species is around 160 in Japan and 900 in the world. Species living on rocks graze on algae, and those burrowing into soft sediments ingest particulate organic matter.
131 Sponge Date unknown / UMUT
Melo aethiopica
Date unknown / Locality unknown / Dried specimen / UMUT
132 *Cymbium amphora*
Date unknown / Locality unknown / Dried specimen / UMUT
133 *Pinctada margaritifera*

Date unknown / Locality unknown / Dried specimen / UMUT

134 Coral Specimen Collection

Corals are a group of cnidarian animals which are phylogenetically close to sea anemones and jelly fish. Living animals secrete a hard skeleton of calcium carbonate and form coral reefs. The majority of corals consist of genetically identical, small individuals called polyps. Each polyp feeds on planktons and also uses organic nutrients produced by photosynthetic symbiotic microalgae (zooxanthellae). Because the algae need intensive sunlight, the distribution of corals is restricted to shallow-water tropical to subtropical areas.

135 Coral Specimen Collection

136 Fur Seal (IMTAb_YZ0000002)

Date unknown / Skeleton specimen / Formerly a collection of Yaizu Fishery High School, Shizuoka / UMUT

137 Globe (1/28000000)

Revised in 1963 / Unsei Watanabe / Japan / Formerly a collection of the Tokyo Institute of Technology / UMUT

138 Mushroom Model Collection

Specimen Formerly a collection of Yaizu Fishery High School, Shizuoka

139 Mushroom Model Collection

Early 20th Century / Made in Japan / Paper clay, color on wood, in glass case / Department of Medicine, UMUT

140 A Species of Monitor (IMTAd_YZ0000001)

1933 / Stuffed specimen / Formerly a collection of Yaizu Fishery High School, Shizuoka / UMUT

Japanese Sawshark (IMTAe_YZ0000001)

Date unknown / Stuffed specimen / Formerly a collection of Yaizu Fishery High School, Shizuoka / UMUT

Sturgeon (IMTAe_YZ0000002)

Date unknown / Stuffed specimen / Formerly a collection of Yaizu Fishery High School, Shizuoka / UMUT

141 Mineral Collection

Mineralogy, geology and mining engineering were initiated in Japan with the courses given by German mining engineer Karl Schenk at the Kaisei Gakko established in 1873. Among the collections inherited from the Kaisei Gakko, numerous mineral and rock specimens imported from the Krantz Mineralien-Kontor company in Germany still remain today. Thanks to the active specimen collection conducted by Schenk's disciples Bunjiro Koto and Tsunashiro Wada, as well as their successors, an enormous collection of minerals, rocks and ores was formed at the University of Tokyo.

142 Pyrite (IMTC_UT0000070)

Date unknown / Navajun, La Rioja (Spain) / Private collection

Pyrite is a sulfide mineral of iron. It shows hexahedron, octahedron and pentagonal dodecahedron forms, as well as their possible combinations. It can be found commonly in various regions, and its value as a raw material is low. That is because it cannot constitute a source of iron, due to the difficulty in separating iron and sulfur completely. In the past, pyrite was extracted in order to produce sulfuric acid, but as sulfur is now derived from petroleum, its volume of extraction considerably decreased. As the relationship between its diverse forms and the conditions of its formation is still unclear, it has sufficient value as a research material.

143 Stibnite (IMTC_UT0000073)

Date unknown / Ichinokawa mine (Ehime) / Wakabayashi collection, Department of Mineralogy, UMUT

Stibnite is a sulfide mineral of antimony of which it is a resource mineral. From the ancient times, antimony was used for cosmetics, and was an important resource materials in modern industry. However, since its toxicity came to be known, it was seldom used. Stibnite is named after the Greek "stimmi" or "stib". The Ichinokawa mine is a famous locality of stibnite.

144 Quartz (Japanese Twin) (IMTC_UT0000075)

Date unknown / Otome mine (Yamanashi) / Wakabayashi collection, Department of Mineralogy, UMUT

Two separate crystals of the same kind which developed attached to each other with a crystallographically significant orientation relation are called twins. Among quartz many kinds of twins are known and the type of the present specimen, characterized by its V-like form, is called "Japanese Twin." The angle between the two crystals is 84o34'. This type of twin is often observed in quartz specimens from the Otome mine (Yamanashi Pref.). Quartz usually shows a hexagonal prism form but the Japanese twin shows a hexagonal plate form, the reason of which is not clarified yet.

145 Jasper (IMTC_UT0000071)
Date unknown / Tamayucho (Shimane) / Private collection
Jasper is an aggregate of fine grains of quartz. It shows various colors and patterns due to impurities. Some beautiful ones belong to the category of gems.

146-147 Globe (1/8000000)
Before 1937 / Belgium / Color on monochrome map / Donated by the Belgium Government / Formerly a collection of the University of Tokyo Library / UMUT
The Great Kanto Earthquake of September 1923 reduced the University Library's 500 000 books to ashes. Help was offered from within and outside Japan, and the Belgium government sent books and donations. To commemorate this, a large-sized globe was ordered to Belgium. However, when the globe was actually delivered in 1937, it was amidst the Second Sino-Japanese War, and Belgium being a member of the United Allies, the friendship following the disaster was already a thing of the past. That may explain why the globe was delivered with an uncolored map. According to the records of the time, "Taking into account the taste of the Japanese, the Belgian Geography Association left the map white, without coloring it." The coloring was done in Japan, from 1939 on.

148-149 Statue of a Chief (gowe) (70.2001.27.543)
19th Century / Niha Population / North-west of Nias Island, Lidano Lahömi basin (Indonesia) / Stone / Former collection of the Barbier-Mueller Museum, Geneva / Quai Branly Museum
The Nias Island, located off Sumatra in the Indian Ocean, offered to the 19th-Century visitor a stunning variety of stone sculptures, in regard to its relatively modest dimension (120 km long and 40 km wide). The architecture, wood and stone sculpture can be stylistically divided into three regions: the North, the Centre and the South. The hilly topography of the island has long maintained these three regions relatively isolated from each other. In the end of the 19th Century, Italian geographer Elio Modigliani brought back from his trip photographs of a village chief from the Southern region. These help us better understand how headdresses and ornaments were worn, and how they were combined. This stone sculpture represents a village chief seated with his hands joined on his chest. He bears his rank's insignia: a nifatali twisted necklace initially in gold, a long earring and a bracelet around his wrists. The ornamented headband girding his forehead under a conic headdress indicates that the finery required by his rank as well as the related feasts have well been realized. The gesture attracting our attention to his chest signifies that this chief is "like a mother to his people", counterbalancing the virility expressed by his phallus. The Niha society is governed by feasts of merit which allow for the acquisition of honorary titles. This social process stimulated competition among chiefs, and was a powerful catalyst for the realization of sculptures. Realized in honor of a chief, this gowe perpetuates the presence and the spirit of an ancestor.

150 Mushroom Model Collection
Early 20th Century / Made in Japan / Paper clay, color on wood, in glass case / Department of Medicine, UMUT

151 Human Anatomical Model (IMTE_MD0001076, 1077, 1078)
Color on plaster / Late 19th Century / Germany / Department of Medicine, UMUT

152 Human Anatomical Model (IMTE_MD0000184)
Date unknown / Color on plaster / Department of Medicine, UMUT

153 Death Mask of Gakutaro Osawa (IMTJ_UT0000192) 1892 / Plaster / UMUT

154 Lion Head of Urakami Cathedral (IMTC_UT0000100)
Discovered on May 13, 1946 / Nagasaki / Department of Petrology and Mineral Deposits, UMUT
Collected during the investigations on the damages of the atomic bomb conducted in 1945-46 by Takeo Watanabe, Professor at the University of Tokyo. This item was originally said to be the guardian lion-dog of the Hiroshima Gokoku Shrine, but it was actually the lion head decorating the arch surmounting the entrance of the Urakami Cathedral in Nagasaki. On the top of the head made of andesite, we can observe glass

resulting from the sudden cooling of black minerals contained in the andesite and which melted with the radiation. It was located about 500 meters away from the centre of the explosion.

155 Mineral Collection

156 Models of the Celebrated Large Historical Diamonds Collection (IMTC_UT0000101)
1873-1893 / Made in England / Glass / Department of Petrology and Mineral Deposits, UMUT
It is estimated that the production of diamond replicas has begun in the 18th Century, in relationship with the considerable development of Natural History at that time. Diamonds have long belonged to the realm of the unknown, and as such they were the object of numerous legends. Among the numerous existing diamonds, the fact of possessing the famous ones basically signifies the domination of mankind over nature. Although they are replicas, these specimens are particularly precious in that they embody the diamonds' actual condition in the 19th Century.

158 Representative Mineral Specimens of the U.S.S.R. (IMTC_UT0000001)
Date unknown / Department of Mineralogy, UMUT
A present from the Mission of the U.S.S.R. Academy of Sciences presided by Dr. Keldysh, when he visited the University of Tokyo in 1964. This collection consists of 16 specimens: Fluorite, Ultramarine, Topaz, Opal, Rhodonite, Talc, Chalcedony, Agate, Cinnabar, Apophyllite, Realgar, Eudialyte, Pyrite, Sulphur, Astrophyllite and Orpiment.

159 Mineral Sample Specimen Collection from Europe (IMTE_MK0000299)
Date unknown / In wooden box / Formerly a collection of the Medical Library, The University of Tokyo / UMUT Miyake collection
About these specimens collected by Hiizu Miyake's father, Gonsai (1817-1868), an attached note mentions: "These were once passed on to Siebold, but they were later returned to our country." Upon Siebold's second visit to Japan, Gonsai asked h(IMTo identify the minerals he had collected. However, once in Europe, Siebold did not comply with Gonsai's demand to return them. It is Hiizu who, on his father's demand, met Siebold during his stay in Europe, and brought back these specimens to Japan. However, these being mainly minerals from Bohemia, they are thought to be specimens received in replacement of those originally collected by Gonsai, which are now lost.

Mineral Collection from Russia (IMTC_UT0000002)
Data unknown / Department of Mineralogy, UMUT
This mineral sample collection is believed to have been brought back from Russia by Prof. Kotora Jinbo, who became a Professor of mineralogy at the University of Tokyo in 1896.

160-161 Quartz (IMTC_UT0000023)
Date unknown / Ishikawa district (Fukushima) / Formerly a collection of the Department of Mining, The University of Tokyo / UMUT

Fossil of coxa, Elephantidae (replica) (IMTD_UT0005012) UMUT

Fossil of Ammonite (IMTD_UT0005015) UMUT

162 Globe (IMTE_MK0000410)
1800s / Berlin (Germany) / Paper and coloring on wooden frame, compass needles in metal / Formerly a collection of the Medical Library, the University of Tokyo / UMUT Miyake collection
This globe was manufactured in Germany and succeeded into today by Miyake family, which one of the famous doctoral families in Japan.

Telescope (IMTE_MK0000411)
19th Century / Made by P. Dorffel / Berlin (Germany) / Brass / Formerly a collection of the Medical Library, the University of Tokyo / UMUT Miyake collection
One of the telescopes made in Europe and brought to Japan. This one was made in Germany and brought today by Miyake family, which one of the members is Miyaka Hiizu (1848-1938), the first dean of the college of medicine, the Imperial University, and with other members, widely known in the history of medicine.

163 Yayoi Pottery
Yayoi period / Yayoi district (Tokyo) / Replica made by Kyoto Kagaku / Anthropology and Prehistory department, UMUT
This is the "type" or "first" specimen of the Yayoi style pot, discovered in 1884 at Yayoi

district, at or adjacent to the University of Tokyo campus. It was, then, of an unknown "new" pottery style intermediate between the seemingly older "stone age" (Jomon) and more recent pottery, and was nicknamed the "Yayoi pottery". Active discussion in the latest 1800s led to the recognition of a new pottery style, culture, and prehistoric age, all of which came to be named after the initially found pot. Hence the Yayoi style pottery, the Yayoi culture, and the Yayoi period (500 B.C. to 300 A.D.).

164 Stone Rods Collection

Late to Latest Jomon period / Polished stone implement / Anthropology and Prehistory department, UMUT

A type of polished stone implement, considered to have been used in ritual ceremony, made from the Early to Latest Jomon period. Distributed widely from western Japan to the northern island Hokkaido, centered in northeastern Japan, they are classified according to morphology, size and manufacturing technique. The exhibited items were collected around 100 years ago through donation or from survey of the University of Tokyo anthropological staff, and are thought to belong to the Late and Latest Jomon period (circa 4000 to 3000 BP).

Tridacna Shell Adze Collection

Date unknown / Rota Island (Northern Mariana Islands) / Shell implement / Anthropology and Prehistory department, UMUT

The exhibited materials are adze bits made from large Tridacna shells. These were collected from Rota Island of the Northern Mariana Islands as a part of the 1910-30s anthropological expeditions of the University of Tokyo. The adze bits are made from the shell bodies, a style characteristic of the Mariana Islands region. In prehistoric Japan, a different style of shell adzes is known from the southernmost Ryukyu Islands.

165 Stone Implement Reference Collection

Late 19th Century / In steel case with glass window / Originally from Peabody Museum, Massachusetts, U.S.A / Ex-collection of the Department of Zoology, Faculty of Science / Anthropology and Prehistory department, UMUT

E. S. Morse, the first professor of zoology at the University of Tokyo, was a specialist of brachiopods, but is also well known for conducting the first scientific excavation in Japan at the Omori shell mounds. Morse returned to the U.S. in 1879, but had strongly recommended that the University found a museum, which was realized as the Science Museum at Kanda. The materials exhibited here are examples of prehistoric implements, which are thought to be those donated to the University of Tokyo, in the 1880s, through Morse's efforts.

166-167 35 mm Film Projector (IMTF0000306)

Date unknown / Fuji Machinery Co., Ltd. / Metal / Formerly a collection of Lecture Hall, Department of Law, the University of Tokyo / UMUT

233 Photograph of a Micromollusc

Printed in 2012 / Ink-jet print / Specimen collected in 1994 / Off Kakeroma Island (Kagoshima), 310 m deep / Dried specimen original size: 4mm / UMUT

This micromollusc is a species of *Cochlespira* (Mollusca: Gastropoda). Shells attain various sizes from species to species. The largest species, *Tridacna gigas*, reaches more than 1m, while the smallest group, the family Omalogyridae, is 0.5mm in maximum diameter. Small species below 1cm are very common. Shells smaller than 4mm are conventionally called "micromolluscs."

234 Stairway

236-237 Bird Stuffed Specimen Collection

Yamashina Institute of Ornithology Collection (Entrusted to UMUT)

The Yamashina Institute of Ornithology is the only institute specialized in ornithology (science of birds) in Japan. The institute emanates from a private collection room established in Marquess Yoshimaro Yamashina's mansion in 1932, and now preserves the best ornithological collection in Japan, both in quality and quantity. Here are stored about 350 specimens deposited from the Yamashina collection, which includes the specimen originally possessed by the Biological Laboratory of the Imperial Household or donated to the Emperor. These old elegant specimens reveal the highest level of art of the taxidermists and craftsmen in those days.

Bird Stuffed Specimen Collection

1930-1980's / Collected by Keikichi and Masao Oita / Formerly Oita Wild Bird Museum Collection / UMUT

Specimens donated from the Oita Wild Bird Museum in Gifu prefecture. Keikichi Oita and Masao Oita ran that private museum for education and bird preservation, but the museum closed in 2008. We received about 300 specimen, most of which are birds, and store part of the collection here.

238-239 Bird Stuffed Specimen Collection
Yamashina Institute of Ornithology Collection (Entrusted to UMUT)

240 Phoenix (Akazasa, Brown Laced) (IMTAc_WK000079)
Kochi Prefecture / Special National Natural Treasure

241 Silver Pheasant (IMTAc_YS0000634)
1982 / Yamashina Institute of Ornithology (Entrusted to UMUT)

242 Emperor Penguin (IMTAc_YS0000376)
Date unknown / Formerly a collection of the Biological Laboratory of the Imperial Household / Yamashina Institute of Ornithology (Entrusted to UMUT)

Bird Stuffed Specimen Collection)
1930-1980's / Collected by Keikichi and Masao Oita / Formerly Oita Wild Bird Museum Collection / UMUT

243 White-tailed Sea Eagle (IMTAc_YS0000470)
Date unknown / Made by Fukuji Sakamoto / Formerly a collection of the Biological Laboratory of the Imperial Household / Yamashina Institute of Ornithology (Entrusted to UMUT)

245 Balance (IMTF0000070)
Date unknown / Made by Sadakichi Moriya / Tokyo / UMUT

247 Microscope (IMTF0000046 Date unknown / R. Fuess/ Berlin / UMUT

248 Sextant (IMTF0000066) Date unknown / K. Hattori & Co. Ltd / Tokyo / UMUT

249 Precision Measuring Instrument (IMTF0000122)
Date unknown / Manufactured by Imai Seiki Seisakusho / UMUT

251 Bust of Juntaro Takahashi (IMTJ_UT0000101)
1921 / Sculpted by Kozaburo Takeishi / Bronze / Formerly a collection of the Department of Pharmacology, The University of Tokyo / UMUT
Juntaro Takahashi (1855-1920) was the first professor at the Department of Pharmacology. He studied in Germany from 1882 to 1885, following his graduation from the Department of Medicine. It is upon his return that Pharmacology came to be taught independently from Medicine. Kozaburo Takeishi (1877-1963) was a Niigata-born sculptor active in the Ministry of Education Bunten exhibition. At the Department of Sculpture of the Tokyo School of Art he studied under Moriyoshi Naganuma, and upon graduating studied at the Belgian Royal School of Art in Brussels from 1901 to 1909. From 1911 on, he exhibited at the Bunten, and gained authority as a portrait sculptor.

Bust of Juntaro Takahashi (IMTJ_UT0000100)
1921 / Sculpted by Kozaburo Takeishi / Plaster / Formerly a collection of the Department of Pharmacology, The University of Tokyo / UMUT

252 Model of Akishinodera Temple Kanshitsuzo Statue. Re-sculpted work of Kanshitsuzo (Dry Lacquered Wooden Buddha Statue), Akishinodera
2008 / Toshimasa Kikuchi / Cypress, lacquer / Private collection
The Akishinodera Temple Kanshitsuzo (dry lacquered wooden Buddha statue) made at the end of the Tempyo period (8th Century), has a unique structure even among other similar dry lacquered statues. In this period where works made out of a single piece of wood appear, the structure exemplifies the relationship between the dry lacquer technique and wood sculpture.

254 Speed Governor (IMTF0000005)
Date unknown / Schaeffer & Budenberg / Buckau-Magdeburg (Germany), Manchester (England) / Brass, steel, wooden base / UMUT

Flexible Joint (IMTF0000002)
Date unknown / Von L. Schröder / Darmstadt (Germany) / Brass, steel, wooden base / UMUT

255 Material Testing Machine (IMTF0000147)
Date unknown / Manufactured by Elliott Brothers Co. / London (England) / Brass, steel, wooden base / Formerly a collection of the Department of Material Engineering, the University of Tokyo / UMUT
Instrument for testing the strength of materials. Various types of such instruments

have been known as a basis for each field of engineering, including architecture, civil engineering and mechanical engineering. This machine is for tensile tests, and comprises spring scales as a measuring device and a scissor jack as a burdening device, thus subjecting the tested material to a relatively heavy load.

256 Grand Prize Diploma of Japan-British Exhibition Awarded to the Tokyo Imperial University College of Engineering
1910 / London (England) / Printed on paper / Formerly a collection of Faculty of Engineering, the University of Tokyo / UMUT
The Japan-British Exhibition was an international exhibition held in Shepherds Bush, London, from May 14 to October 29, 1910 after the conclusion of the Anglo-Japanese Alliance, and on vast grounds covering 22,550 m2. The honorary president from Japan was Prince Fushimi Sadanaru. After its victory against Russia in 1905, Japan was eager to catch up with the great Western powers, and this exhibition was regarded as a good opportunity for Japan to appeal its modernized nation and to promote its export extension policy toward the U.K.

257 Chart of Faculty of Forensics Medicine

258 Mathematical Model: Minimal Surface with a One-Parameter Family of Parabolas
Late 19th Century to early 20th Century / Made by Martin Schilling Co. (Germany) / Plaster, iron / Formerly a collection of the Department of Mathematics, Tokyo Imperial University / Department of Mathematics, the University of Tokyo
The surface of this model is parametrized as above with a mean curvature of $H=0$. This surface is an example of a minimal surface containing a one-parameter family of parabolas. It is either a general form of Catalan's minimal surface, or such a surface with an additional helicoidal form.

259 Horn of Greater Kudu (*Tragelaphus strepsiceros*)
Date unknown / Dried specimen / Private collection
The kudu is a large antelope that inhabits areas of eastern and southern Africa. It stands at about 150 cm in height and weighs as much as 300 kg. Its horns twist around more than twice as they develop and can reach sizes as great as 180 cm. For the most part, the kudu lives in open forest and rarely inhabits grasslands. Only males have evolved horns for sexual selection.

260-263 Cabinet of curiosities

264 Hydrodynamic Form (Based on Propeller Model)
2011 / Toshimasa Kikuchi / Cypress, lacquer, brass / Private collection
Sculpture imaged from a propeller model for experiments in fluid dynamics. The functional demand to transform air resistance into propulsion created this perfected "form." This model can be seen as the product of experimental science, but at the same time resembles some natural "forms," such as the seed of a plant.

265 Photographer unknown Portrait of a Japanese Woman Sitting
Date unknown / Original size: H9.1 x W5.9cm / From the Christian Polak collection
Photographer unknown Portrait of a Japanese Woman
1862 / Original size: H16.1 x W12.5cm / From the Christian Polak collection

266-267 Kuichi Uchida Portrait of the Meiji Emperor
1875 / Original size: H27.0 x W18.5 / From the UMUT collection
Kuichi Uchida Portrait of the Meiji Empress
1875 / Original size: H27.0 x W18.5cm / From the UMUT collection
Photographer unknown Portrait of a Japanese Woman
Date unknown / Original size: H13.9 x W9.3cm / From the Christian Polak collection
Photographer unknown Portrait of Two Japanese Women
Date unknown / Original size: H13.5 x W9.0cm / From the Christian Polak collection

268-269 Disderi Portrait of Akitake Tokugawa
1867 / Original size: H8.8 x W5.5cm / From the Christian Polak collection
Frederick Sutton Portrait of Shogun Yoshinobu Tokugawa
May 1867 / Osaka / Original size: H9.7 x W7.7cm / From the Christian Polak collection
Photographer unknown Portrait of a Japanese Woman with a Shamisen
Date unknown / Original size: H13.9 x W9.3cm / From the Christian Polak collection
Photographer unknown Portrait of a Japanese Woman with a Parasol
Date unknown / Original size: H8.7 x W5.5cm / From the Christian Polak collection

270-271 Nadar Portrait of Sukekuni Kawazu
1864 / Paris / Original size: H29.9 x W21.3cm / From the Christian Polak collection

Photographer unknown Portrait of Tadayoshi Shimazu
Date unknown / Original size: H14.7 x W10.4cm / From the Christian Polak collection
Photographer unknown Portrait of a Japanese Woman with a Parasol
Date unknown / Original size: H14.4 x W9.3cm / From the Christian Polak collection
Photographer unknown Portrait of a Japanese Woman
Date unknown / Original size: H13.9 x W9.3cm / From the Christian Polak collection

272 Genpei Akasegawa Greater Japan Zero-Yen Note
1969 / Paper printed on both sides / Private collection
Banknotes are printed material. What has been mass-produced in tens of thousands, in hundreds of thousands of exemplars without a single deviation, is precisely the authentic banknote. Genpei Akasegawa's Model 1000-yen Note, which precisely exploited this ordinary mechanism of our times based on such an economic system of currency, was deemed to violate the Act on Control of Imitation of Currency and Securities, and Akasegawa was indicted in 1964, and held guilty by the Supreme Court in 1970. This item is a Real Zero-yen Note conceived by Akasegawa, and is a particularly appropriate piece of Conceptual Art. Nowadays, the Zero-yen Note is sold at high prices on the art market, which recognizes it as a work of art having a high historical value.
Inflation Banknotes from the Weimar Republic
1914-1923 / Germany / Printed on paper / UMUT
In Germany, an excessive inflation occurred in less than ten years after World War Ⅰ. These notes are from the one mark banknote in 1914 to the five hundred million mark note in 1923.

273 Coin Collection

274 Miyake Family Collection from the End of the Edo Period
19th Century to early 20th Century / Formerly a collection of the Medical Library, the University of Tokyo / UMUT
Former collection of the Miyake family, a famous family of doctors around the central figure of Hiizu Miyake (1848-1938). Hiizu Miyake was the first Doctor in Medical Science in Japan, and served successively as a professor of Medicine at the University of Tokyo, dean of the same faculty, honorary professor and member of the House of Peers. In 1863 and until the following year, he participated to the Tokugawa shogunate's Embassy to Europe as its youngest attendant. Numerous medical instruments and scientific specimens he bought then remain to us. He continued to acquire numerous foreign products upon his travels overseas.

275 Six-Chambered Revolver (IMTE0000415)
19th Century / Belgium / Steel, grip in wood, wooden case
Revolver initially owned by Hiizu Miyake's father Gonsai (1817-1868). The barrel bearing the engraving "9th Year, 1457, Tokyo", it is assumed that it was registered in Tokyo in 1876 (Meiji 9).

276 Miyake Family Collection from the End of the Edo Period
19th Century to early 20th Century / Formerly a collection of the Medical Library, the University of Tokyo / UMUT

277 Counterweight (IMTE0000355) Metal, wooden case

278 Gastroscope (IMTE0000336) Metal, leather case

279 Set of Surgical Instruments (IMTE0000307)
Made by Leiter / Wien (Austria) / Metal, tortoiseshell, leather case

280 Forms and Forces

281 Perspective Model (Series VIII, nr.4) (IMTG0000108)
1882 / Made by Haale Co. / Germany / Plaster, iron wire / Formerly a collection of Department of Mathematics, Tokyo Imperial University / Graduate School of Mathematical Science, the University of Tokyo

282 Gourd Date unknown / Taiwan / Private collection

283 Ostrich Egg Date unknown / Private collection
The ostrich's egg is the largest among living birds, but in proportion to the bird's body mass, it is the smallest in size. It is a bird which evolved by producing numerous eggs, which do not match its body dimensions.

284 Mathematics Educational Material
Late 19th Century / Wood / Formerly a collection of the Department of Mathematics, Tokyo Imperial University / Graduate School of Mathematical Science, The University of Tokyo

285 Precious coral
Data unknown / Locality unknown / Dried specimen / Department of Research, UMUT
The word coral indicates cnidarian amimals forming hard skeletons. Among them species used for jewelry are called precious corals. Their animal consists of a colony of numerous polyps and covers a hard skeleton of calcium carbonate. The skeleton with branches is beautifully red in color. Most species inhabit rocky bottoms in deep water.
286 Model of Native Gold (IMTC_UT0000093)
19th Century / Krantz Mineralien-Kontor / Germany / Plaster / Kranz collection, Department of Petrology and Mineral Deposits, UMUT
Replica of a gold nugget excavated in Russia in 1842, which was the biggest (about 40 kg) at the time. It was produced by Dr. Krantz, a famous mineral specimen company in Germany. Such a replica has high academic value as a three-dimensional record of the excavated original form.
287 Schorl (NaFe3Al6(BO3) 3Si6O18(OH)4) (IMTC_UT0000096)
Date unknown / Golconda Mine (Brazil) / Department of Mineralogy, UMUT
The realm of Nature is abundant in astonishing forms. Above all, the world of minerals is remarkable for its diversity. Schorl belongs to the tourmaline group, and is black-colored because it contains iron. It is mainly formed in granite. As it polarizes when subjected to heat or pressure, it is commonly called "electric stone" in Japanese. Among the tourmaline group, the species which contain lithium bear diverse and beautiful colors, and they are prized as precious stones.
288 Calcite Nodule (IMTC_UT0000095)
Date unknown / Tsukinuno mine (Yamagata) / Department of Mineralogy, UMUT
Round-shaped calcite found in bentonite. Its particles are often found in hot springs as a result of deposition.
289 Black-Lacquered Foiled Box with Cover
Yuan Dynasty (China), 14th century / Wood, lacquer / Giichi Tanaka collection
Foiled box shaped as a flower with ten petals. It is supposed that a plate made of hawksbill turtle shell fitted on top of the cover. The seam of the box as well as the contour of the petals are outlined with lead. This piece was produced in the transition period from the Northern Song Dynasty lacquerware, with characteristic neat forms and beautiful lacquered colors, to a more exuberant and powerful expression.
290 Crested Bombonieres Bestowed by the Imperial Family
Silver / Giichi Tanaka collection
"Bonbonnière", meaning "small candy box" in French, is a piece of metal craft offered to guests on such special occasions as weddings and religious ceremonies. In Japan, the Imperial Family introduced this European tradition, and started using bombonieres bearing the Imperial chrysanthemum crest. In the international society of the 1880s, silver had the same value as paper money, and silverware was frequently used for royal diplomacy in the West. Bombonieres were mainly offered by the Imperial Family on the occasion of a birth ceremony, the fifth birthday of a child, the coming-of-age ceremony, the investiture of the Crown Prince, a wedding, as well as for commemorating a visit abroad or for celebrating longevity. They were also distributed upon luncheons with foreign royals or ambassadors. Among the seven exhibited items, the three offered to celebrate the wedding of the Meiji Emperor's daughter bear a mention by Rintaro Tanaka specifying on which occasion they were given. As for the three items respectively offered on the occasion of the State ceremony for the Taisho Emperor, the investiture and the coming-of-age ceremony of the Showa Emperor, the letters of invitation to each banquet sent to Fuji Tanaka by Yoshinao Hatano, the Minister of the Imperial Household, remain.
291 Shaped as a Six-Petal Incense Box with Mandarin Duck Motif
November 27, 1916 / Gift for the investiture of Prince Hirohito Michinomiya (the Showa Emperor) / Engraved "Pure silver made by Kobayashi"
292 Untitled (Photographs of The Crown Prince's Palace)
1909 / Photographed by Kazumasa Ogawa / 154 sheets, 2 cases, collotype printing, black leather case, wooden slip case / Giichi Tanaka collection
Photographs of the Crown Prince's Palace (current Akasaka Palace, national treasure), upon its completion. It is a piece of European palace architecture in the neo-baroque style designed by Tokuma Katayama and constructed to house Prince Yoshihito (the

Taisho Emperor). The structure is in steel and bricks, and the outer wall made of stone. The best architects, artists and engineers were reunited for this construction, which took ten years to complete. The photographs cover the whole building: the façade, interior decoration, bathroom, corridors and underground facilities. Rintaro Tanaka was granted these photographs for his effort in building leading-edge heating equipment. The Imperial Household Archives hold a gelatin-silver print photo album with almost the same dimensions. It is assumed that there were two types of albums, a luxurious version with a white leather folding case and a normal one with a black leather folding case, dedicated to the Imperial Family. Kazumasa Ogawa produced the collotype version and distributed it to interested parties. Ogawa was named a court artist in 1910, the first time for a photographer.

293 Mathematical Model Collection Made by the German Martin Schilling Company
Late 19th Century to early 20th Century / Plaster (replica based on the collection of the Department of Mathematics, the University of Tokyo, formerly a collection of the Department of Mathematics, Tokyo Imperial University) / UMUT
These models can divided into those relating to algebraic geometry, to differential geometry and to the theory of complex functions. These are not conceptual models, but minute models based on actual numeric calculation. The modeling of complex surfaces being difficult, such a realization with a high degree of precision is most precious. The original models preserved at the Department of Mathematics of the Faculty of Science were imported in the early 20th Century by Professor Senkichi Nakagawa, and were actually used in the Department's courses. The Martin Schilling Company ceased its model production in 1932 due to the decrease of demand.

294-296 Mathematical Models

297 Sievert's Surface--surface of constant positive curvature (IMTG_UM0000114)
Plaster (Replica) / 2011 / Made by Toshimasa KIKUCHI / UMUT

298 Generalized helicoid of constant positive curvature (IMTG_UM0000070)
Plaster (Replica) / 2011 / Made by Toshimasa KIKUCHI / UMUT

299 Mathematical Model: Surface with a singular point (IMTG_UM0000113)
Plaster (Replica) / 2011 / Made by Toshimasa KIKUCHI / UMUT

304 Collection of Ship Models Made by Takesada Tokugawa and Hydrodynamic Instruments (IMTF0000132)
Early Showa Period / Wood / Formerly a collection of the Department of Nautical Engineering, the University of Tokyo / UMUT
Models built to experimentally prove the relation of a boat's form and its propulsion resistance. As a belonging of Takesada Tokugawa (1888-1957), the models were passed down to the Department of Nautical Engineering in the Engineering School at Tokyo Imperial University. Takesada, a naval technical lieutenant general, chief of the Naval Technical Institute and professor at the Imperial University of Tokyo, was born the second eldest son of Akitake Tokugawa, the last lord of the Mito domain, and later became the founder of the Matsudo Tokugawa branch.

305 Collection of Ship Models Made by Takesada Tokugawa and Hydrodynamic Instruments (IMTF0000132)
Early Showa Period / Wood / Formerly a collection of the Department of Nautical Engineering, the University of Tokyo / UMUT

306 The Namio Egami Collection
Namio Egami (1906–2002), who was professor emeritus at the University of Tokyo, is best known among the public for his influential theory regarding the formation of the ancient State in Japan. However, his interests were wide-ranging and his important contributions to the study of history and archaeology over a period of 70 years were not restricted to Japan, and extended to nearly all parts of the Eurasian continent. Being an authority on human science pertaining to Eurasia, he was awarded the Order of Cultural Merit in 1991. Not only was he a great researcher, but he was also a matchless collector of archaeological, historical and artistic objects. On repeated occasions of fieldwork during his long academic career, including extensive archaeological excavations in Iraq and Iran in the 1950-1960s, he eventually constituted an enormous antique collection. The specimens on display are part of this collection, which serves as an invaluable teaching-research material for the history of ancient Eurasia.

307 Necklaces of Ancient Persia (IMTH_EG_000045)
Namio Egami collection, Department of Archaeology, UMUT

Early ornaments were manufactured with materials available in nature, notably precious stones, shells, and ivory. From the Neolithic period onwards, artificial materials started being used for making ornaments. The newer materials included ceramic and metal. Faience was one of these, and was unique and popular in the ancient civilizations of the Middle East. It was a type of ceramic, often greenish–blue in color, made not from clay but from crushed quartz or sand. The development of technology with respect to faience appears to have lead to the production of glass later.

Ring-Shaped Bronze Objects
8th to 6th centuries B.C. / Iran / Department of Archaeology, UMUT
These bronze objects belong to the late Iron Age culture of North Iran. They have an open ring shape with tapering ends, around which fine incised decorations are visible. A suggestion has been made that these unique objects were used as ingots for metal trade.

308 Necklaces of Ancient Persia (IMTH_EG_000046)
Namio Egami collection, Department of Archaeology, UMUT

309 Glass Vessels of the Ancient Orient (IMTH_EG_0000034, 0000036, 0000037)
After the Roman period / Department of Archaeology, UMUT
The manufacturing of glass dates back to the third millennium B.C. in the Ancient Mesopotamia. While the main products at the beginning were small beads and seals, glass vessels were introduced sometime around 1500 B.C. Glass-blowing technique was developed in the Roman period. This technique enabled the production of a far larger variety of glass objects. Some of the glass vessels made in the Sasanian period (3–7th Century A.D.) were quite thick, and are thought to have suited long-distance trade on land. One such example is a glass bowl with facets in relief (right). It is the same type of bowl as that stored at the Shosoin Imperial Treasure House.

310 Glass Vessels of the Ancient Orient (IMTH_UT_000016)
After the Roman period / Department of Archaeology, UMUT

311 Hump-Backed Bull Pottery (IMTH_EG_000033)
Late 2nd to early 1st millennia B.C. / Northwestern Iran / Namio Egami collection, Department of Archaeology, UMUT
This pottery vessel represents a hump-backed bull, a popular domesticated animal in northwestern Iran. While the humped back appears relatively natural, the spouted mouth is evidently stylized, reflecting the earlier tradition of spouted vessel production. The pierced ears suggest the original use of metal earrings. Such vessels are usually found intact in human graves. They were probably utilized during funeral rituals, but their exact functions and utilization remain unknown.

312 Ancient Pottery from Iran (IMTH_UT_000022)
Late prehistoric (3rd millennium B.C.) to historic age / Northern Iran / Department of Archaeology, UMUT
Historic pottery collected by the University of Tokyo archaeological mission in northern Iran. Animal motifs were quite common for the ancient Iranian pottery in general. The long spout of the dull brown Iron Age jars probably imitates a bird's bill.

313 Bronze Swords from Ancient Iran (IMTH_UT_000010、000014)
Late 2nd to early 1st millennia B.C. / Northern Iran / Department of Archaeology, UMUT
These bronze swords were collected by the University of Tokyo archaeological mission in the late 1950s. The manufacturing of bronze objects continued in this period of the Early Iron Age. Their symbolic value is indicated by the fact that many of them have been discovered as funerary items.

314-315 Japanese Cock Stuffed Specimen Collection
Late 1900's / Collected by Nosan Corporation / Formerly Wakeikan collection / UMUT
These stuffed specimens were donated from Wakeikan, a Japanese cock collection room within Nosan Corporation. In Japan, many local breeds had been made for various purposes such as cockfight, eating or to compete with crowing. Chabo, or Japanese bantam, is a Japanese unique breed made for companion animals. Many of these Japanese cocks are protected as National Natural Treasures.

316 Scarlet *Minohiki* (IMTAc_WK000033)
Aichi and Shizuoka Prefectures / National Natural Treasure

317 Beard Fowl (IMTAc_WK000021) Sado Island, Niigata Prefecture

318 *Minohiki chabo* (Ohiki) (IMTAc_WK000041)
Kochi Prefecture / National Natural Treasure

319 Silver Laced (Japanese Bantam) (IMTAc_WK000074)
320 Koeyoshi (IMTAc_WK000030) Akita prefecture / National Natural Treasure
321 Real-Black (Japanese Bantam) (IMTAc_WK000053)
322 White-Peony (Japanese Bantam) (IMTAc_WK000046)
323 Black-Tailed White (Japanese Bantam) (IMTAc_WK000076)
324 Brown Partridge-Tailed (IMTAc_WK000018)
Kochi Prefecture / National Natural Treasure
Silver Banded (Japanese Bantam) (IMTAc_WK000050)
325 White Fighting Cock (IMTAc_WK000049) National Natural Treasure
326 Giant Insects in Shigeru Eda Collection
Mainly late 20th Century / Dried specimens / UMUT
Shigeru Eda (1930-2008) was a great collector of strange and beautiful insects throughout the world. Some gigantic and curious items from his collection donated to the UMUT are shown here. In the current earth environment, the body size of insects is limited by their body structure, body mass and respiratory system, as well as by predators. Among these specimens, *Attacus atlas* is the largest Lepidoptera in Japan. Leaf insects are a miracle of evolution.
327 *Eurycnema versirubra* (IMTAf_ED0001294)
Attacus atlas (Male) (IMTAf_ED0001274)
Attacus atlas (Female) (IMTAf_ED0001275)
Attacus atlas (Cocoon) (IMTAf_ED0001276)

328 *Salvazana mirabilis* (IMTAf_ED0001271)
Ayuthia spectabile (IMTAf_ED0001272)
Salvazana mirabilis (IMTAf_ED0001273)
329 *Phyllium celebicum* (IMTAf_ED0001287)
Phyllium celebicum (IMTAf_ED0001288)
Phyllium giganteum (IMTAf_ED0001289)
330-331 Ornithopter by Leonardo da Vinci (1/20 Scale Model)
1996 / Giovanni Sacchi / Color on wood, canvas, string, leather / Private collection
The National Museum of Science and Technology Leonardo da Vinci in Milano is known for exhibiting replica models of mechanisms and machines conceived by Leonardo. Most of them were made by woodworker Giovanni Sacchi. The fundamental form of a machine with its wings spread open has a rich legacy, from Sir George Cayley's flying machine and the Lilienthal brothers' glider to Vladimir Tatlin's "Letatlin." But it is Leonardo who took the first step in mankind's dream of flying.
Leonardo da Vinci Codex on the Flight of Birds, from the Biblioteca Reale in Turin (*Il Codice sul Volo degli Uccelli*, nella Biblioteca Reale di Torino)
ca. March-April, 1505 / 1979 / Iwanami Shoten Publishers / Japanese edition of facsimile / UMUT
This manuscript contains, from the recto of the 18th leaf to the verso of the 4th leaf, the Codex on the Flight of Birds. According to the interpretation of Ichiro Tani, Ken'ichi Ono and Yasuhiro Saito, by "bird", Leonardo designates both winged animals and the flying machines he conceived. For the latter, the machines named "giant bird", there were three types: a spring-powered one, one with a bat-like wing, and one where the flyer lies prone. During his three years in Florence from 1503 on, he observed in detail the flight of raptors such as horned owls, and attempted to apply the results of his observations to his plans for human flight.
332-333 Albino Indian Peacock (Tail), Copper Pheasant (Tail), Indian Peacock (tertiary), Common Pheasant (Tail), Raptor (Primary Remex)
2012 / Ink-jet print / Feather specimens formerly from the Oita Wild Bird Museum collection / UMUT
Greater Bird of Paradise
2012 / Ink-jet print / Stuffed specimen formerly from the Oita Wild Bird Museum collection / UMUT
334 Bird Stuffed Specimen Collection
1930-1980's / Collected by Keikichi and Masao Oita / Formerly Oita Wild Bird Museum Collection / UMUT
335 Ural Owl (IMTAc_OT0001280)
Formerly Oita Wild Bird Museum Collection / UMUT
336 Paradise Flycatcher (Male, Female) (IMTAc_OT0001053, 0001054)

Formerly Oita Wild Bird Museum Collection / UMUT
337 Ruddy Kingfisher (IMTAc_OT0001153)
Formerly Oita Wild Bird Museum Collection / UMUT
338 Banded Pitta (IMTAc_OT0001177)
Formerly Oita Wild Bird Museum Collection / UMUT
339 Crested Serpent Eagle (IMTAc_OT0001308)
Formerly Oita Wild Bird Museum Collection / UMUT
340 Goosander (IMTAc_OT0001071)
Formerly Oita Wild Bird Museum Collection / UMUT
341 Great Spotted Woodpecker (IMTAc_OT0001116)
Formerly Oita Wild Bird Museum Collection / UMUT
342 Copper Pheasant (Male) (IMTAc_OT0001044)
1981 / Formerly Oita Wild Bird Museum Collection / UMUT
343 Little Egret (Winter Plumage) (IMTAc_OT0001073)
1980 / Formerly Oita Wild Bird Museum Collection / UMUT
344 Grey-Faced Buzzard (Female) (IMTAc_OT0001036)
Formerly Oita Wild Bird Museum Collection / UMUT
345 Collared Scops Owl (IMTAc_OT0001127)
Formerly Oita Wild Bird Museum Collection / UMUT
346 Japanese Crested Ibis (IMTAc_YZ0000001)
Date unknown / Stuffed specimen / Formerly a collection of Yaizu Fishery High School, Shizuoka / UMUT
The specimen had been preserved in a high school since 1965, and is estimated to be an individual of the Japanese population that still survived back then.
347 *Phoenicopterus* sp. (Species of Flamingo) (IMTAc_OT0001201)
Formerly Oita Wild Bird Museum Collection / UMUT
348 Grey Heron (IMTAc_OT0001070)
Formerly Oita Wild Bird Museum Collection / UMUT
349 Roseate Spoonbill (IMTAc_OT0001187)
Formerly Oita Wild Bird Museum Collection / UMUT
350 Japanese Pheasant (Male) (IMTAc_OT0001040)
1980 / Formerly Oita Wild Bird Museum Collection / UMUT
351 Mandarin Duck (Male) (IMTAc_OT0001128)
Formerly Oita Wild Bird Museum Collection / UMUT
352 Capercaillie (IMTAc_OT0001200)
Formerly Oita Wild Bird Museum Collection / UMUT
353 Sarus Crane (IMTAc_UT0000018) Date unknown / UMUT
354 Donsongoni (Hunter's Harp)
Bougouni area, Republic of Mali / Bambara people (Mandé peoples)
355 Bala (Balafon) Kita area, Republic of Mali / Malinke people (Mandé peoples)
356 Kesekese (Caxixi) Republic of Mali / Mandé peoples
357 Tama (talking drum) Nioro area, Republic of Mali / Mandé Peoples
358 Ngoni (African-Lute) Futa Toro area, Republic of Mali / Mandé Peoples
359 Karinyan (Güiro) Republic of Mali / Mandé peoples
360 Set of Drawing Instruments (IMTF0000129)
February 1876 / Workshop of the Imperial College of Engineering, Japanese Ministry of Engineering / Brass, ivory, ebonite, wood, in cedar wood case / Formerly a collection of the Engineering Design Room of the Industrial Engineering Department, the University of Tokyo / UMUT
This engineering heritage is most precious in that it proves the existence of an inscription original to the Imperial College of Engineering (ICE) of the Ministry of Works in the first years of Meiji. The ICE was an education institution for engineering established within the Ministry of Works in order to introduce engineering technology from the West. With British teachers such as Henry Dyer, the College opened in August 1873 with an integrated training in engineering spanning over 6 years (including a preparatory course, a specialized course and a practical course of 2 years each). The inscription "February 1876" corresponds to the period when the first promotion entered the specialized course, after having completed the preparatory course. Technical drawing was a fundamental discipline for the preparatory and specialized courses in mechanical and naval engineering.

361 Measure (IMTF0000131)
Date unknown / Metal / Formerly a collection of the Department of Geography, the University of Tokyo / UMUT
362 Mechanical Model Collection of the School of Engineering
1870's / Brass, iron, wooden base / Formerly a collection of the Department of Mechanical Engineering, the University of Tokyo / UMUT
These models show the mechanism of the transmission of force and movement through various elements, and were used in teaching mechanical engineering since the early Meiji era. They illustrate fundamental mechanisms such as the screw mechanism, the linkage mechanism, the composition of a lever, the crank and slider, and the gear mechanism. Some of them bear a metallic plate dating from 1874 and 1875: they are some of the oldest heritage in Western industrial mechanical engineering remaining in Japan. All of them coming from England, it is thought that they were ordered from his home country along with books related to engineering by Henry Dyer, who was appointed Principal of the Imperial College of Engineering in 1873.
363 Triangular Cam (IMTF0000004)
364 Differential Screw (IMTF0000011)
365 Differential Gear (IMTF0000013)
366 Quick Return Mechanism (IMTF0000032)
Von L. Schröder / Darmstadt (Germany) / Metal
367 Mechanical Model (IMTF0000023) 1875
368 Four Link Mechanism (IMTF0000015)
369 Worm Wheel Mechanism (IMTF0000018)
370-371 Ratchet (IMTF0000010)
Mechanical Model (IMTF0000024)
372 Metal Cutting Early Return Mechanism (IMTF0000014)
373 Mechanical Model (IMTF0000031)
374 Swash Plate Mechanism (IMTF0000012) Gustav Voigt / Berlin (Germany)
375 Inner Spur Gear Mechanism (IMTF0000019)
376 External Flat Gear Mechanism (IMTF0000007)
377 Reciprocating Motion Mechanism using a Gear (IMTF_UT0000009)
378 Heart-Profiled Cam Mechanism (IMTF0000001) 1874
This type of fundamental mechanism consisting of a driver (cam) rotating around a fixed axis and a follower accompanying it regularly is called a "cam." This model served for educational purposes in the Imperial College of Engineering in the early Meiji era. The distance from the cam's center of rotation to its edge being proportional to the driver's angle, the follower's velocity is stable when the cam's angular velocity is.
379 Thermometer
380 Typewriter (IMTF0000116) Royal Co., Ltd.
381 Legume of Matchbox Bean (Entada rheedei) (IMTB_UT0000596)
2005 / Myanmar / Dried specimen / Formerly Herbarium, University of Tokyo collection / UMUT
The matchbox bean is distributed in tropical and subtropical areas throughout Asia, Australia and Africa. In Japan, species of this genus occur in coastal regions in the islands south of Kyushu. The plants are large, evergreen, woody lianas. The stem can reach 30 cm in diameter. The woody legume (fruit) can reach 1 m in length. The seeds float and sometimes drift ashore on the coast. The Japanese name, *Modama*, designates the seed of seaweed because people believed they came from the sea. The seeds are russet, round, 5 cm in diameter and are used for handicrafts.
382-383 Plants of the Himalayas
There are flowers of various colors blooming in alpine Himalayas. With long winters and short summers, rain almost everyday during blooming, and strong ultraviolet rays on sunny days the alpine regions in the Himalayas are a severe environment for plants. Against this severe environment, the Himalayan plants show elegant beauty for growing, blooming, and fruiting. Since the 1960s, the University of Tokyo has been sending many botanical expedition teams to the Himalayas to clarify the diversity of plants in this region. As a result, we have acquired a wide knowledge of the Himalayan plants, and numerous specimens, which are the basis for research, have been accumulated.
John James Laforest Audubon *The Birds of America* (Reproduction)
1830-1839 / Published by the Author / London (England) / Color lithography on paper /

UMUT
Born in the West Indies as the natural son of a British Navy officer, Audubon (1785-1851) travelled all over the Northern American continent by foot, capturing birds, making stuffed specimens, and producing this reference book, considered as the pinnacle of 19th-Century illustrated ornithological books. In this reproduction in color lithography, the birds of America are figured in actual size in their environment. Each of these drawings contain numerous birds with differing sex and age, drawn so as to show their original characteristics, making this book a field guide. It is also an ecological encyclopedia presenting the birds' natural environment and characteristic behavior. The background shows surrounding plants as well as what the birds feed on, with their scientific name meticulously indicated. It is said that Matthew Calbraith Perry offered a full set of this book to the Tokugawa shogun, but the existence of such as copy has not yet been verified.

Bird Stuffed Specimen Collection
1930-1980's / Collected by Keikichi and Masao Oita / Formerly Oita Wild Bird Museum Collection / UMUT

384 *Swertia multicaulis* D. Don (Gentianaceae) (IMTB_HM0000383)

385 *Aconitum hookeri* Stapf (Ranunculaceae) (IMTB_HM0000384)

386 John James Laforest Audubon
Blue Jay / *The Birds of America* (Reproduction) / UMUT

387 Noble Rhubarb, Sikkim Rhubarb, *Rheum nobile* (Polygonaceae) (IMTB_HM0000595)
2012 / Jaljale Himal (Eastern Nepal) / Dried specimen / Department of Botany, UMUT
A perennial herb, it occurs in alpine regions of the Himalayas. It grows more than 1.5 m tall above the ground and nearly 1 m across at the ground, contrasting with many dwarf plants surrounding it. It has green leaves on the lower part of the stem near the ground, and creamy white, semitransparent ones (bracts) on its upper part.

388 Pharmacological Specimen Collection
Department of Pharmaceutical Science, UMUT
The UMUT holds a collection of pharmacological specimens amounting to 15000 varieties, mainly consisting of research specimens of the Faculty of Pharmaceutical Science of the University of Tokyo. The existence of numerous specimens dating back to the prewar era gives us an insight into the evolution of research in pharmaceutical science. Pharmacology comprises natural materials such as plants, animals and minerals which serve as base materials for medical and pharmaceutical purposes. In order to clearly establish the constituents of medicine comprising pharmacological elements, historical pharmacological samples are indispensable. Amidst the exhaustion of natural resources, pharmacological specimens become hard to obtain, and those accumulated in research institutes constitute an important resource for the development of new medicine as well as for research in leading-edge science.

389 John James Laforest Audubon
Pileated Woodpecker, Racoon Grape/ *The Birds of America* (Reproduction) / UMUT

390 John James Laforest Audubon
American Crow, Black Walnut / *The Birds of America* (Reproduction) / UMUT

391 Titan Arum (*Amorphophallus titanum*) (IMTB_UT0000597)
Printed in 2012 / Photographed by Jin Murata in July 2010 / Ink-jet print / Cultivated by the Botanical Gardens, Faculty of Science, the University of Tokyo / Original size: 1.5m / UMUT
Rare species endemic to Sumatra, Indonesia. It blossomed for the first time in Japan at the Koishikawa Botanical Gardens in 1991 and this specimen, photographed in 2010, is only the sixth successful blooming of the species in Japan. When the large spathe opens, a strong smell like that of rotting meat comes from the spadix, attracting pollinator insects. The small male and female flowers gather at the base of the spadix covered by the spathe. The height of the spadix can reach 3.3m and the diameter of the spathe can reach 1.3m. The name Titan signifies "giant," and the Japanese name *Shokudaiookonnyaku* means candlestick giant arum, because of the shape of the flower.

392 Seeds and Fruits Collection
1950's / Established by Dr. Sachiko Kurosawa / Dried specimen / Department of Botany, UMUT
Seeds may have a hard coat to protect the juvenile plant inside from drought, water and

predators. Even though the plant cannot move by itself, they can be carried far away from the parent plant. Fruits are designed to release seeds by hanging and splitting open. Fruits and seeds may have wings or long hairs like cotton that carry them with the wind. Corky tissues make them light so that they can be carried by wind or to aid in floating on ocean currents and in rivers and streams.

393 Tropical Fruits Collection
Taisho to early Showa era / Dried specimen / Department of Botany, UMUT
During the Taisho and early Showa eras, researches for plant resources in SE Asia and Oceania were vigorously conducted by the University of Tokyo. These fruits are a part of the collection from that period.

394 Pandanus boninensis Warburg (Pandanaceae) (IMTB_UT0000438)
1923 / Ogasawara (Bonin) Islands
It grows in Ogasawara Islands. The Japanese name "Takonoki" means "tree of octopus", comparing the lower stem with many aerial prop roots to an octopus.

395 Mangrove Palm; Nipa Palm: Nypa fruticans Wurmb (Palmae) (IMTB_UT0000434)
Date unknown / Locality unknown
This palm is well known as a mangrove plant. It is distributed widely in the tropics and subtropics in Asia and Oceania. It occurs also in Ishigaki and Iriomote Islands, south Japan.

396 Black Coral (Antipatharia)
Purchased in 2011 / Philippines / Dried specimen / UMUT
As in the "form" of trees, adapting to the ecological environment of the ocean floor, the animal's body fans out as it grows. Even its fine cilia branch out following the same principle. Specimens like this, which show little damage, are most difficult to acquire.

397 Sago Palm: Metroxylon sagu Rottb. (Palmae) (IMTB_UT0000435)
1915 / Truck Islands (Chuuk Islands)
This palm is distributed in SE Asia and Oceania and widely cultivated. We can get starch from the trunk of this palm, and make food such as pan cakes.

398 Sea Poison Tree; Fish Poison Tree: Barringtonia asiatica (L.) Kurz. (Lecythidaceae) (IMTB_UT0000437)
Date unknown / Locality unknown
It is distributed in the Ryukyus, Taiwan, SE Asia, and Oceania. Box fruits (fruits of Sea Poison Tree) are used as a fish poison. The Japanese name "*Goban-no-ashi*" means "legs of a *go* board", because the shape of the fruits look alike.

399 Greater Bird of Paradise (IMTAc_UT0000024)
Late 20th Century / Stuffed specimen / UMUT
This species distributes in the New-Guinean rain forest. The male birds of the *Paradisaea* have various ornamental feathers, displaying it to attract females. Birds of Paradise were commonly hunted because their feathers were used for decoration, but they are now protected by international treaties.

COLONNADE

天秤
年代未詳／守谷定吉造／東京
／東京大学総合研究博物館研究部所蔵
Balance
Date unknown / Made by Sadakichi Moriya
/ Tokyo / UMUT

650
50
100
200
250
300
400
LONDON
展示物にお手を触れないでください。
Please do not touch the exhibits.

DIPLOMA FOR
GRAND PRIZE
JAPAN-BRITISH
EXHIBITION. 1910
AWARDED TO
日英博覽會
SHEPHERDS BUSH, LONDON

指紋 (III)

Please do not touch the exhibits.

Photographer unknown
Portrait of a Japanese Woman
Photographer unknown
Portrait of a Japanese Woman Sitting

撮影
東京大学総合研究博物館研究部所蔵より
the Meiji Empress
From the UMUT collection
明治天皇肖像
東京大学総合研究博物館研究部蔵より
Uchida
the Meiji Emperor
Original size: H27.0 x W18.5 / From the UMUT collection

撮影者未詳
日本人女性肖像
年代未詳／原寸縦 13.9、横 9.3cm ／クリスチャン・ポラック氏所蔵より
Photographer unknown
Portrait of a Japanese Woman
Date unknown / Original size: H13.9 x W9.3cm / From the Christian Polak collection
撮影者未詳
二人の日本人女性肖像
年代未詳／原寸縦 13.5、横 9.0cm ／クリスティアン・ポラック氏
Photographer unknown
Portrait of Two Japanese Women
Date unknown / Original size: H13.5 x W9.0cm / From the Christian

ディスデリ
徳川昭武肖像
1867年／原寸縦 8.8、横 5.5cm／クリスティアン・ポラック氏所蔵より
Disderi
Portrait of Akitake Tokugawa
1867 / Original size: H8.8 x W5.5cm / From the Christian Polak collection
フレデリック・サットン
徳川慶喜肖像
1867年5月／大阪／原寸縦 9.7、横 7.7cm／クリスティアン・ポラック氏所蔵より
Frederick Sutton
Portrait of Shogun Yoshinobu Tokugawa
May 1867 / Osaka / Original size: H9.7 x W7.7cm / From the Christian Polak collection

撮影者未詳
三味線を持つ日本人女性肖像
年代未詳／原寸縦13.9、横9.3cm／クリスティアン・ポラック氏所蔵より
Photographer unknown
Portrait of a Japanese Woman with a Shamisen
Date unknown / Original size: H13.9 x W9.3cm / From the Christian Polak collection
撮影者未詳
日傘を持つ日本人女性肖像
年代未詳／原寸縦8.7、横5.5cm／クリスティアン・ポラック氏所蔵より
Photographer unknown
Portrait of a Japanese Woman with a Parasol

ナダール
河津祐邦肖像
1864年／パリ／原寸縦29.9、横21.3cm／クリスティアン・ポラック氏所蔵より
Nadar
Portrait of Sukekuni Kawazu
1864 / Paris / Original size: H29.9 x W21.3cm / From the Christian Polak collection
撮影者未詳
島津忠義肖像
年代未詳／原寸縦14.7、横10.4cm／クリスティアン・ポラック氏所蔵より
Photographer unknown
Portrait of Tadayoshi Shimazu
Date unknown / Original size: H14.7 x W10.4cm / From the Christian Polak collection

撮影者未詳
日傘を持つ日本人女性肖像
年代未詳／原寸縦14.4、横9.3cm／クリスティアン・ポラック氏所蔵より
Photographer unknown
Portrait of a Japanese Woman with a Parasol
Date unknown / Original size: H14.4 x W9.3cm / From the Christian Polak collection
撮影者未詳
日本人女性肖像
年代未詳／原寸縦13.9、横9.3cm／クリスチャン・ポラック氏所蔵より
Photographer unknown
Portrait of a Japanese Woman
Date unknown / Original size: H13.9 x W9.3cm / From the Christian Polak

FÜNFZIG MILLIONEN MARK
500 MILLIARDEN MARK
100 Milliarden Mark
Hundert Milliarden Mark
100000 MARK
200000 Mark
大日本帝國政府
拾圓
本
零
円
物
RA565658R

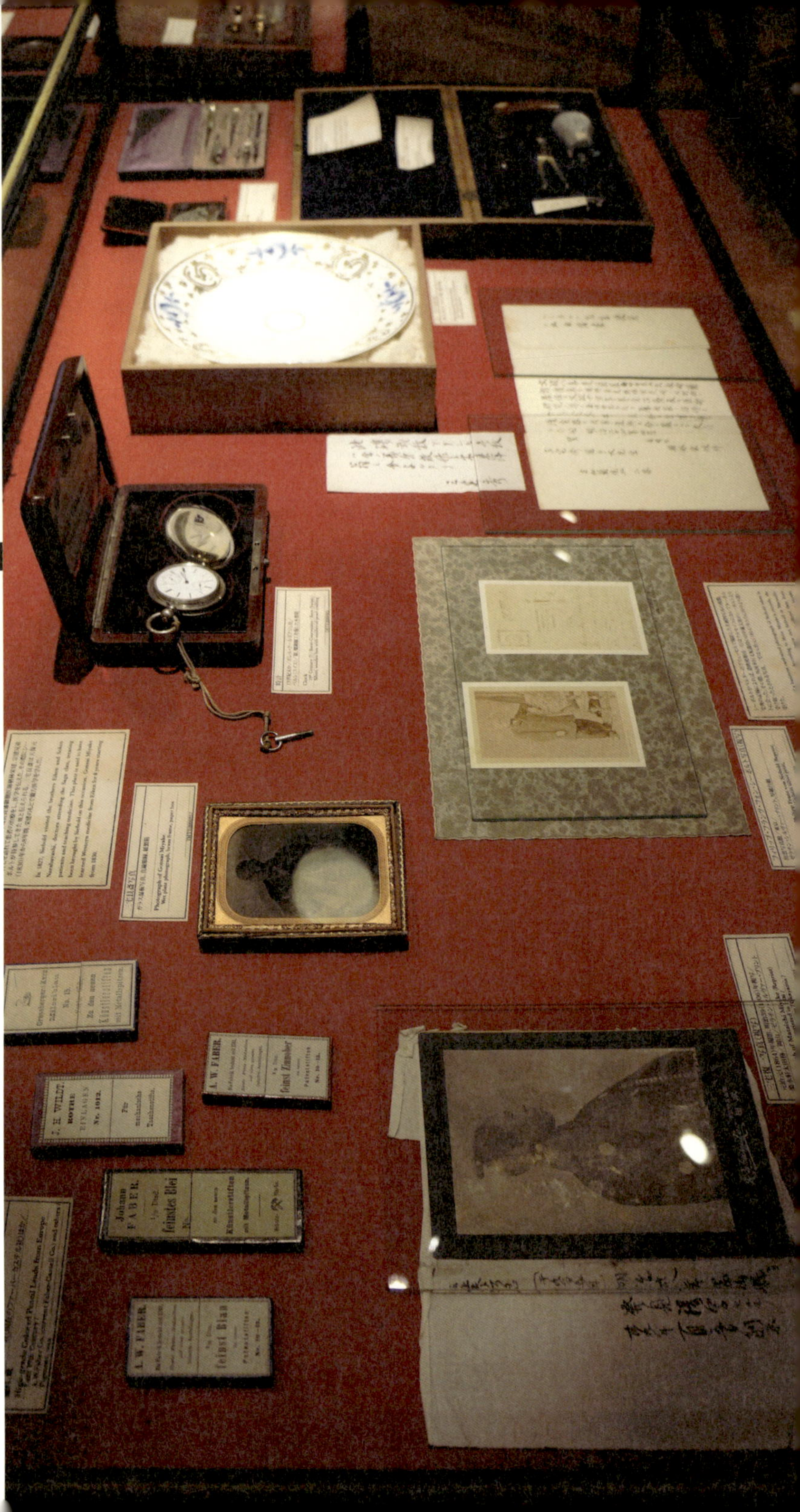
A. W. FABER
feinst Zinnober
J. H. WILDT
Johann FABER
feinstes Blei
A. W. FABER
feinst Blau

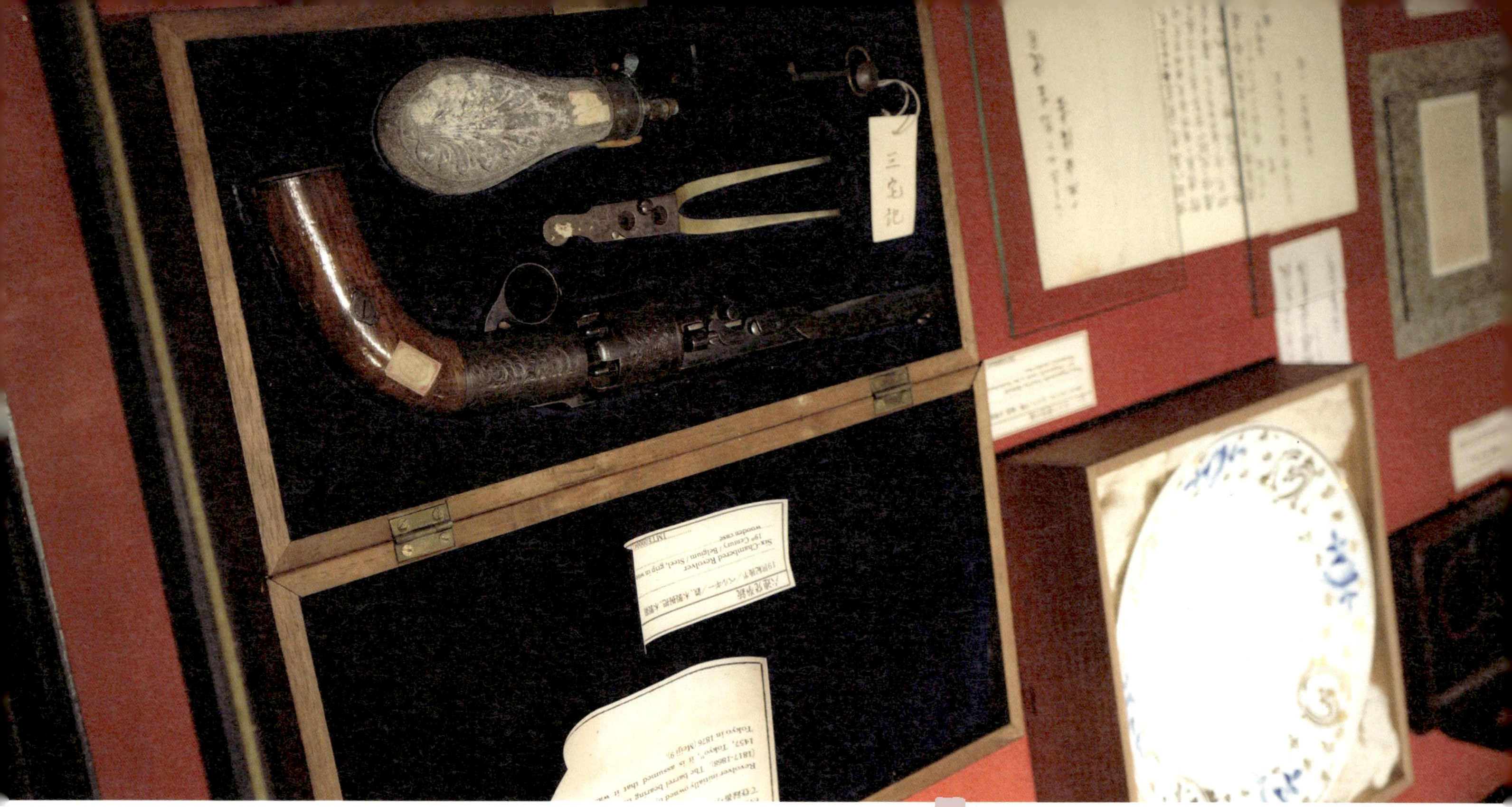

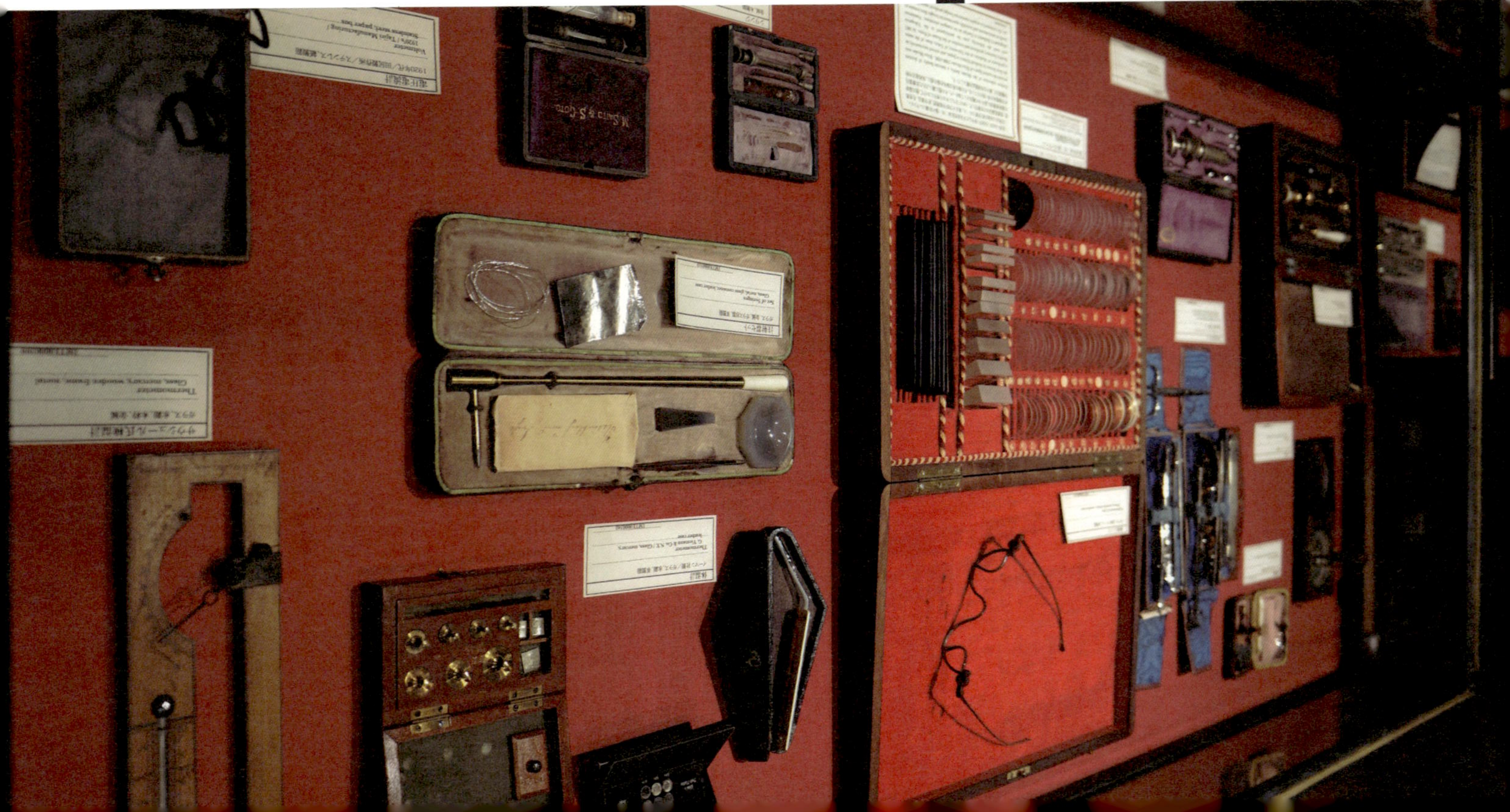

IMTE0000341
Counterweight
Metal, wooden case
IMTE0000355
Abbe's
Test-Plates
Carl Zeiss
Jena
体温計
イーマン社製／ガラス、水銀、革製箱
Thermometer
G. Tiemann & Co. N.Y. / Glass, mercury,
leather case
IMTE0000340

Gastroscope
Metal, leather case

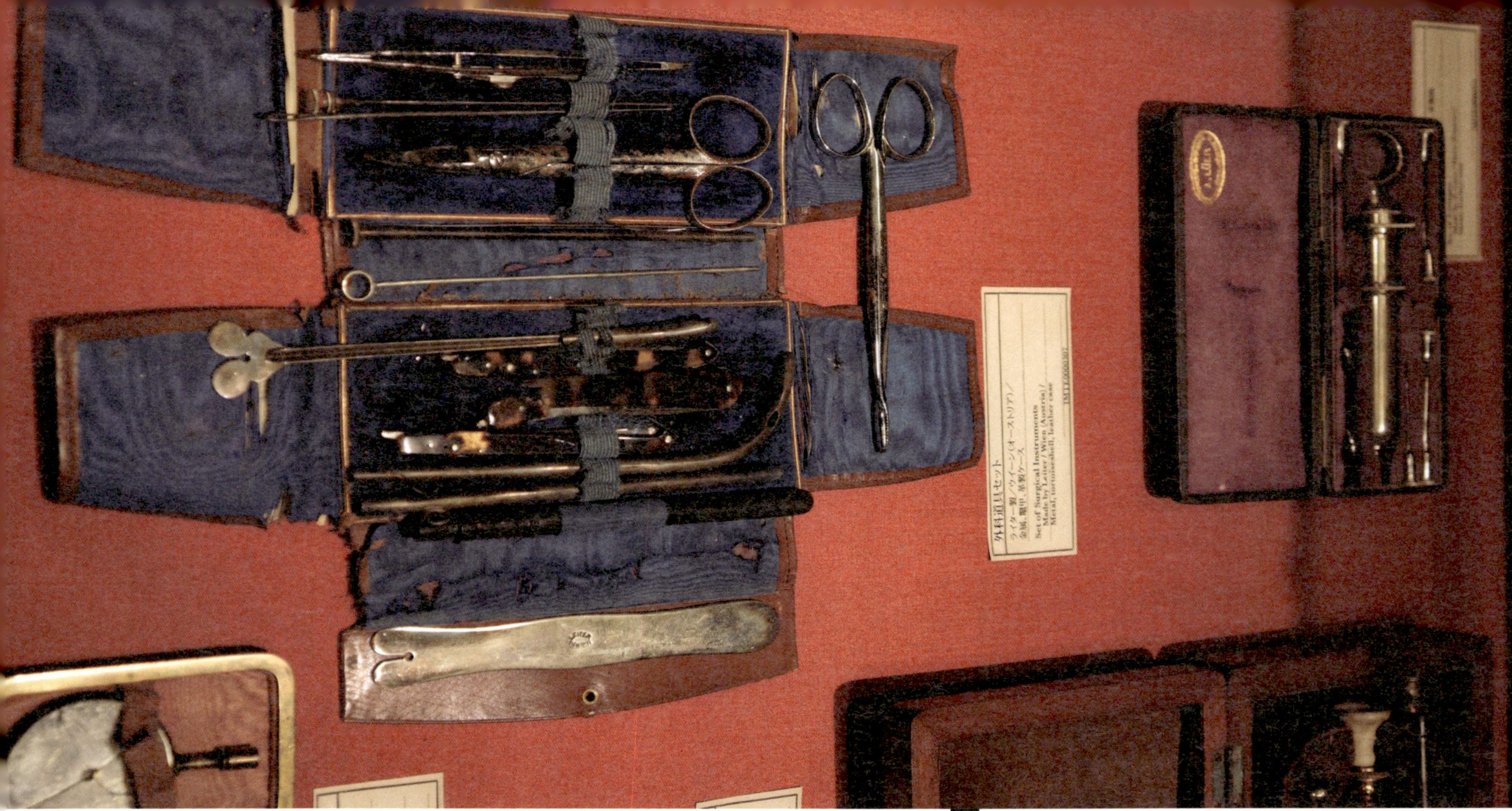
外科道具セット
ライター製／ウィーン(オーストリア)／
金属、鼈甲、革製ケース
Set of Surgical Instruments
Made by Leiter / Wien (Austria) /
Metal, tortoiseshell, leather case

Gourd
Date unknown / Taiwan / Private collection

理学部数学科旧蔵／
Material
ormerly a collection of
atics, Tokyo Imperial
of Mathematical
kyo

年4月29日／北白川宮成久王と房子内親王
御結婚祝宴記念
as a Letter Case
1909 / Gift for the marriage banquet of Prince Naruhisa
Fusako Kitashirakawanomiya
柳箱型
1908(明治41)年4月30日／竹田宮恒久王と昌子内親王
(明治天皇第六女子)御結婚祝宴記念
Shaped as a Yanagibako (Split Willow Wood Box)
April 30, 1908 / Gift for the marriage banquet of Prince Tsunehisa
and Princess Masako Takedanomiya
ボニエール

六葉花弁形香合型
1916(大正5)年11月27日／迪宮裕仁親王
刻印「純銀　小林製」
Shaped as a Six-Petal Incense Box with
November 27, 1916 / Gift for the investit
Michinomiya (the Showa Emperor) /
Engraved "Pure silver made by Kobaya
一當日午前十一時三十分參内
一服装大禮服、正装
一公務又ハ病氣ニ依リ參内致
速ニ式部職ヘ申出ノ事
退出口
東
右

大階段降口

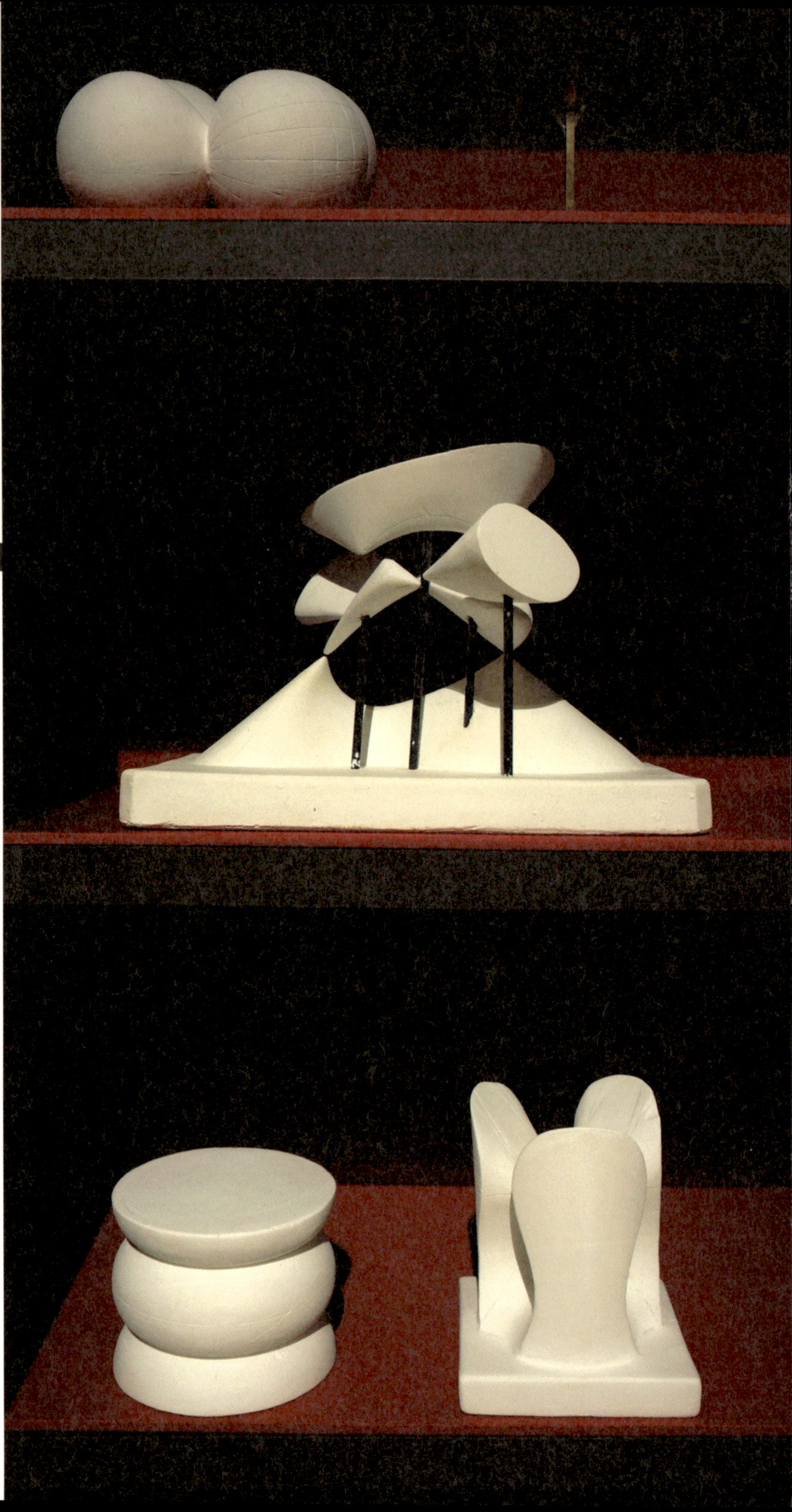

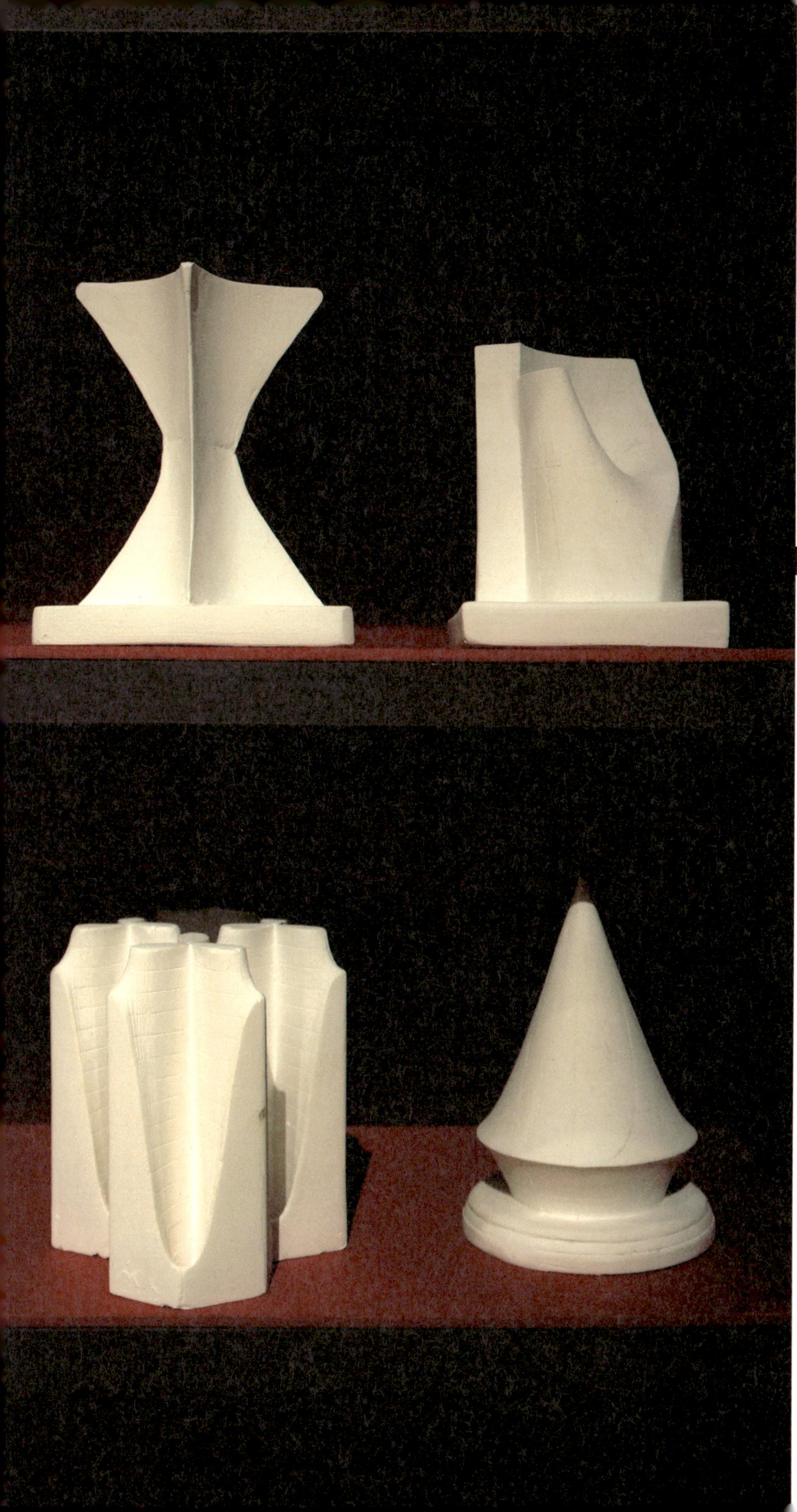

猩々蓑曳(雄)
愛知県, 静岡県／天然記念物
Scarlet Minohiki
Aichi and Shizuoka Prefectures /
National Natural Treasure

新潟県佐渡島
Beard Fowl
Sado Island, Niigata Prefecture

銀笹矮鶏
Silver Laced (Japanese Bantam)
IMTAc_WK000074

声良（雄）
秋田県／天然記念物
Koeyoshi
Akita prefecture / National Natural Treasure

白牡丹矮鶏
White-Peony (Japanese Bantam)

White (Japanese Bantam)
IMTAc_WK000076

Silver Banded (Japanese Bantam)
Brown Partridge-Tailed
Kochi Prefecture / National Natural Treasure

Pseudophyllus titan
アカネクマゼミ
Salvazana mirabilis
アユタヤゼミ
Ayuthia spectabile
モエギクマゼミ
Salvazana mirabilis

センストビナナフシ
Eurynecroscia nigrofasciata
セレベスコノハムシ
Phyllium celebicum
オオコノハムシ
Phyllium giganteum
セレベスコノハムシ
Phyllium celebicum

Leonardo da VINCI
7

UMUT
サンコウチョウ（雄雌）
Paradise Flycatcher (Male, Female)
IMTAc_OT0001053, 0001054

アカショウビン
Ruddy Kingfisher
IMTAc_OT0001153

UMUT

コサギ（冬羽）
1980（昭和55年）
Little Egret (Winter Plumage)
1980

UMUT

Nipponia nippon
静岡県立焼津水産高等学校

キジ（オス）
Japanese Pheasant (Male)
キジ（ヒナ）
Japanese Pheasant

オシドリ（雄）
Mandarin Duck (Male)

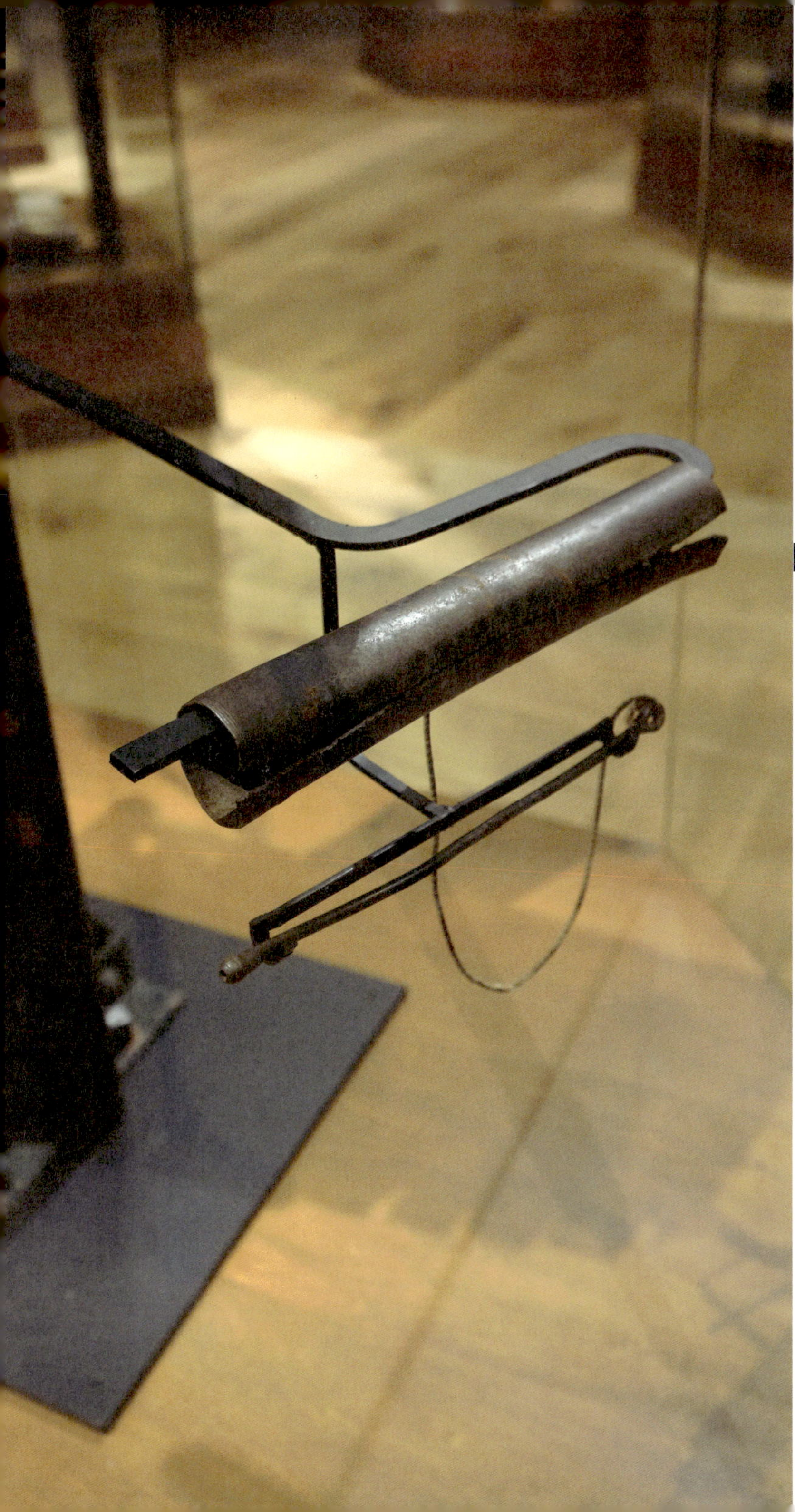

製図器セット
1876(明治9)年2月／大日本工部省工学寮工作所／黄銅、象牙、エボナイト、
木、杉材製ケース入り／東京大学大学院工学系研究科産業機械工学専攻
設計工学研究室旧蔵／総合研究博物館研究部所蔵
Set of Drawing Instruments
February 1876 / Workshop of the Imperial College of Engineering,
Japanese Ministry of Engineering / Brass, ivory, ebonite, wood,
in cedar wood case / Formerly a collection of the Engineering Design
Room of the Industrial Engineering Department,
the University of Tokyo / UMUT
明治最初期の工部省工学寮時代の銘の存在を確認できる極めて貴重な工学遺産。工部省工学寮は、欧米から工業技術を導入するため、工部省の一等寮として設立が認められた工業教育機関である。ヘンリー・ダイヤーら英国人教師を迎え、1873(明治6)年8月に、予科2年、専門科2年、実地科2年の計6年を修業年とした全寮制のエンジニア教育が開始された。「明治九年二月」の年記は第一期生が予科を終え、専門科に進学した頃にあたる。製図学は、機械工学や造船学の予科と専門の基礎学であった。
This engineering heritage is most precious in that it proves the existence of an inscription original to the Imperial College of Engineering (ICE) of the Ministry of Works in the first years of Meiji. The ICE was an education institution for engineering established within the Ministry of Works in order to introduce engineering technology from the West. With British teachers such as Henry Dyer, the College opened in August 1873 with an integrated training in engineering spanning over 6 years (including a preparatory course, a specialized course and a practical course of 2 years each). The inscription "February 1876" corresponds to the period when the first promotion entered the specialized course, after having completed the preparatory course. Technical drawing was a fundamental discipline for the preparatory and specialized courses in mechanical and naval engineering.

44

Differential Screw

Differential Gear

Quick Return Mechanism
Von L. Schröder / Darmstadt (Germany) / Metal

IMPERIAL COLLEGE
OF
ENGINEERING.
TOKEI, 1875.

ウォーム歯車機構
Worm Wheel Mechanism

QUICK

回転斜板機構
グスタフ・フォイクト機械製／ベルリン（ドイツ）
Swash Plate Mechanism
Gustav Voigt / Berlin (Germany)

内かみあい平歯車機構
IMTF0000019
Inner Spur Gear Mechanism

外かみあい平歯車機構
IMTF0000007
External Flat Gear Mechanism

ROYAL

リンドウ科センブリ属の一種
Swertia multicaulis D. Don (Gentianaceae)

ンポウゲ科トリカブト属の一種
hookeri Stapf

Blue Jay.
Male 1. Female 2.3.

A perennial herb, it occurs in alpine regions of the Himalayas. It grows more than 1.5 m tall above the ground and nearly 1 m across at the ground, contrasting with many dwarf plants surrounding it. It has green leaves on the lower part of the stem near the ground, and creamy white, semitransparent ones (bracts) on its upper part.
Univ. Mus., Univ. of Tokyo
セイタカダイオウ(タデ科)
Noble Rhubarb, Sikkim Rhubarb
Rheum nobile (Polygonaceae)

Pileated Woodpecker

American Crow
CORVUS AMERICANUS

インターメディアテク——東京大学学術標本コレクション

発行日　2013年11月1日　初版第1刷

編　者　西野嘉章
編集・発行　東京大学総合研究博物館（UMUT）
〒113-0033 東京都文京区本郷7-3-1
装丁・レイアウト　関岡裕之＋西野嘉章
発　売　株式会社平凡社
〒101-0051 東京都千代田区神田神保町3-29
電話　03-3230-6570（代表）
03-3230-6572（営業）
振替　00180-0-29639
印刷・製本　秋田活版印刷株式会社

ISBN978-4-582-28446-1
NDC分類番号460 A5変型判（21.0cm）　総ページ400
平凡社ホームページ http://www.heibonsha.co.jp/